P9-CFU-924

ALL THE WORLD'S ANIMALS

PRIMATES

ALL THE WORLD'S ANIMALS

PRIMATES

TORSTAR BOOKS
New York · Toronto

CONTRIBUTORS

JDB John D. Baldwin PhD
University of California
Santa Barbara, California
USA

SB-H Sarah Blaffer-Hrdy
Harvard University
Cambridge, Massachusetts
USA

DB-J Douglas Brandon-Jones BSc
British Museum (Natural History)
London

PC-D Pierre Charles-Dominique
Muséum National d'Histoire Naturelle
Brunoy
France

DJC David J. Chivers MA PhD
University of Cambridge
England

THC-B Tim H. Clutton-Brock MA PhD
University of Cambridge
England

JEF John E. Fa
University of Oxford
England

AHH A. H. Harcourt PhD
University of Cambridge
England

CHJ Charles H. Janson
University of Washington
Seattle, Washington
USA

WGK Warren G. Kinzey
City University of New York
N.Y.
USA

HK Hans Kummer
Zürich University
Switzerland

JMacK John MacKinnon DPhil
Bogor, Indonesia

KMacK Kathy MacKinnon MA DPhil
Bogor, Indonesia

RDM Robert D. Martin DPhil FIBiol
University College
London

KM Katharine Milton
University of California
Berkeley, California
USA

RAM Russell A. Mittermeier
State University of New York
Stony Brook, New York
USA

CN Carston Niemitz
Freie Universität Berlin
West Germany

FEP Frank E. Poirier PHD
Ohio State University
Colombus, Ohio
USA

JIP J. I. Pollock BSc PhD
Duke University Primate Center
Durham, North Carolina
USA

AFR Alison F. Richard PhD
Yale University
New Haven, Connecticut
USA

TER Thelma E. Rowell
University of California
Berkeley, California
USA

ABR Anthony B. Rylands PhD
Instituto Nacional de Pesquisas da Amazônia
Manaus
Brazil

BS Barbara Smuts PhD
Harvard University
Cambridge, Massachusetts
USA

TTS Tom T. Struhsaker PhD
New York Zoological Society
Bronx Park, New York
USA

JWT John W. Terborgh
Princeton University
Princeton, New Jersey
USA

PMW Peter M. Waser PhD
Purdue University
West Lafayette, Indiana
USA

AJW Anthony J. Whitten
Centre for Environmental Studies
Medan, Indonesia

RWW Richard W. Wrangham PhD
University of Ann Arbor
Ann Arbor, Michigan
USA

PCW Patricia C. Wright
Duke University Primate Center
Durham, North Carolina
USA

ALL THE WORLD'S ANIMALS PRIMATES

TORSTAR BOOKS INC.
300 E. 42nd Street,
New York, NY 10017

Project Editor: Graham Bateman
Editors: Peter Forbes, Bill MacKeith, Robert Perberdy
Art Editor: Jerry Burman
Picture Research: Linda Proud, Alison Renney
Production: Bob Christie
Design: Chris Munday

Originally planned and produced by:
Equinox (Oxford) Ltd
Mayfield House, 256 Banbury Road
Oxford, OX2 7DH, England

On the cover: Verreaux's sifaka
page 1: Common squirrel monkey
pages 2–3: Long-tailed macaques
pages 4–5: Common langurs
pages 6–7: Slender loris
pages 8–9: Mandrill

Printed in Belgium

Editor
Dr David Macdonald
Animal Behaviour Research Group
University of Oxford
England

Artwork Panels
Priscilla Barrett

Advisory Editors
Dr Tim H. Clutton-Brock
University of Cambridge
England

Dr Russell A. Mittermeier
World Wildlife Fund
USA

Professor Bob D. Martin
University College
London
England

Library of Congress Cataloging in Publication Data

Main entry under title:

Primates.

(All the world's animals)
Bibliography: p.
Includes index.
1. Primates. I. Series.
QL737.P9P676 1985 599.8 85-977
ISBN 0-920269-74-5

CONTENTS

FOREWORD 8

PRIMATES 10

PROSIMIANS 22

LEMURS 24
Ring-tailed lemur, Brown lemur, Sportive lemur, Gentle lemurs, Indri, Sifaka, Aye-aye . . .

BUSH BABIES, LORISES AND POTTOS 36
Bush babies, Angwantibo, Potto, Slender loris, Slow loris . . .

TARSIERS 42

MONKEYS 44

MARMOSETS AND TAMARINS 46
Pygmy marmoset, Tassel-ear marmoset, Emperor tamarin, Lion tamarin, Goeldi's monkey . . .

CAPUCHIN-LIKE MONKEYS 56
Night monkey, Titi monkeys, Squirrel monkey, Capuchins, Sakis, Uakaris, Spider monkeys, Howler monkeys, Woolly monkeys

GUENONS, MACAQUES AND BABOONS 74
Macaques, Baboons, Gelada, Drill, Mandrill, Mangabeys, Guenons, Patas monkey, Talapoin monkey, Allen's swamp monkey . . .

COLOBUS AND LEAF MONKEYS 102
Proboscis monkey, Snub-nosed monkeys, Surelis, Langurs, Leaf monkeys, Colobus monkeys

APES 116

GIBBONS 118
Siamang, Hoolock gibbon, Pileated gibbon, Kloss gibbon . . .

CHIMPANZEES 126
Common chimpanzee, Bonobo

ORANG-UTAN 132

GORILLA 136

TREE SHREWS 144

Pygmy tree shrew, Common tree shrew, Indian tree shrew, Terrestrial tree shrew . . .

FLYING LEMURS 150

Malayan colugo, Philippine colugo

BIBLIOGRAPHY 152

GLOSSARY 153

INDEX 156

ACKNOWLEDGMENTS 160

FOREWORD

The primates—the apes, monkeys and their relatives—are creatures outstanding in the natural world. No other group of animals combines such colorful natural athleticism with considerable intelligence and intricate social relationships. The societies of primates encompass every variant between friendship and feud and are engrossingly interesting, both in themselves and in the comparisons they evoke with the human race.

Primates is an exploration of the animal world that will appeal alike to professional and schoolchild. Here we tell the tales, in words and with superb drawings and photographs, of how gibbons leap between tree branches, how chimpanzees use tools, how male baboons go to war, how mother bush babies rear their young—and much more besides. And interwoven between the descriptions of the many species is the thread of the newest discoveries in modern biology. With the help of the most up-to-date research findings, insights into the lives of primates are revealed whose reality renders our wildest fables dull.

Men and women have long been fascinated by the primates and their world but are often far from being the friends and protectors of these threatened creatures. As the world's tropical rain forests are relentlessly destroyed, so the lives of many primates are extinguished also. Only the renewed and continued protection of primate species, and constant vigilance as to their welfare, will prevent many of the species described in these pages from becoming extinct.

How this book is organized

Animal classification, even for the professional zoologist, can be a thorny problem, and one on which there is scant agreement between experts. This volume has selected the views of many taxonomists but in general follows the classification of Corbet and Hill (see Bibliography) for the arrangement of families and orders.

This volume is structured at a number of levels. First, there is a general essay highlighting common features and main variations of the biology (particularly the body plan), ecology and behavior of the primates and their evolution. Secondly, essays on each of the two suborders of primates (higher and lower) highlight topics of particular interest, but invariably include a distribution map, summary of species or species groupings, description of skull, dentition and unusual skeletal features of representative species and, in many cases, color artwork that enhances the text by illustrating representative species engaged in characteristic activities.

The main text of *Primates*, describing individual families and their species or groups of species, covers details of physical features, distribution, evolutionary history, diet and feeding behavior, as well as their social dynamics and spatial organization, classification, conservation and their relationships with man.

Preceding the discussion of each species or group is a panel of text that provides basic data about size, life span and the like. A map shows its natural distribution, while a scale drawing compares the size of the species with that of a six-foot man. Where there are silhouettes of two animals, they are the largest and smallest representatives of the group. Where the panel covers a large group of species, the species listed as examples are those referred to in the text. For such large groups, the detailed descriptions of species are provided in a separate Table of Species. Unless otherwise stated, dimensions given are for both males and females. Where there is a difference in size between the sexes, the scale drawings show males.

As you read these pages you will marvel as each story unfolds. But as well as relishing the beauty of these primates, you should also be fearful for them. Again and again, authors return to the need to conserve species threatened with extinction and by mismanagement. All the species described in these pages are listed in the Appendices I through III of the Convention on International Trade in Endangered Species of Wild Flora and Fauna (CITES). The *Red Data Book* of the International Union for the Conservation of Nature and Natural Resources lists 42 species of primates at risk. In *Primates*, the following symbols are used to show the status accorded to species by IUCN at the time of going to press. [E] = Endangered—in danger of extinction unless causal factors are modified (these may include habitat destruction and direct exploitation by man). [V] = Vulnerable—likely to become endangered in the near future. [R] = Rare, but neither endangered nor vulnerable at present. [I] = Indeterminate—insufficient information available, but known to be in one of the above categories. [?] = Suspected, but not definitely known to fall into one of the above categories. The symbol [*] indicates entire species, genera or families, in addition to those listed in the *Red Data Book*, that are listed by CITES. Some species and subspecies that have [EX] or probably have [EX?] become extinct in the past 100 years are also indicated.

PRIMATES

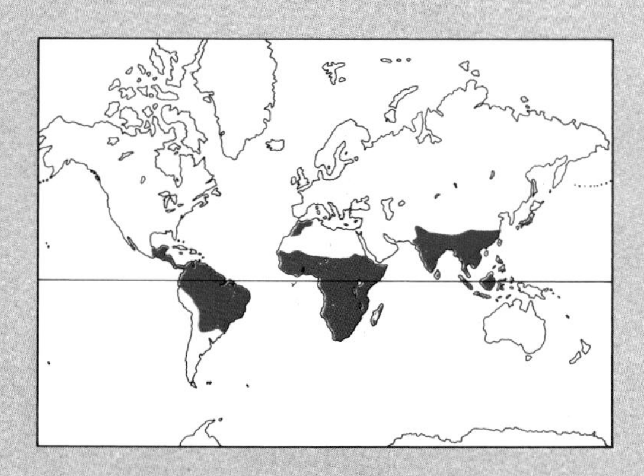

ORDER: PRIMATES
Eleven families; 52 genera; 181 species.

Prosimians (lower primates)
Suborder: Prosimii.

Lemurs
Family Lemuridae
Ten species in 4 genera.
Includes **Ring-tailed lemur** (*Lemur catta*), **Brown lemur** (*L. fulvus*).

Dwarf and mouse lemurs
Family Cheirogaleidae
Seven species in 4 genera.
Includes **Greater dwarf lemur** (*Cheirogaleus major*).

Indri and sifakas
Family Indriidae
Four species in 3 genera.
Includes **indri** (*Indri indri*).

Aye-aye
Family Daubentoniidae
One species, *Daubentonia madagascariensis.*

Bush babies, pottos and lorises
Family Lorisidae
Ten species in 5 genera.
Includes **Dwarf bush baby** (*Galago demidovii*), **Lesser bush baby** (*Galago senegalensis*).

Tarsiers
Family Tarsiidae
Three species in one genus.
Includes **Western tarsier** (*Tarsius bancanus*).

Monkeys and apes (higher primates)
Suborder: Anthropoidea.

Capuchin-like monkeys
Family Cebidae
Thirty species in 11 genera.

To many people typical primates are the acrobatic monkeys or ponderous gorilla, and most recognize that man himself is a primate. Technically primates are an order—a taxonomic category on the same level as the carnivores, rodents, cetaceans (whales) etc. The order Primates contains a far wider array of animals than just monkeys and gorillas. In addition to the capuchin-like monkeys of South and Central America (family Cebidae) and the Old World monkeys of Africa and Asia (family Cercopithecidae), the primates include four families of lemurs from Madagascar, the bush babies and lorises of Africa and Asia, and the tarsiers of Southeast Asia (all prosimians), the marmosets and tamarins of South America, the gibbons and orang-utan and the African apes, and all fossil and living men.

The prosimians, or "lower primates," are less man-like, or show less advanced primate evolutionary trends, than the anthropoids, the "higher primates" which comprise the monkeys, apes and man. Prosimians tend to have longer snouts, a more developed sense of smell, and smaller brains; anthropoids are short-faced, with a highly developed sense of vision, and large brains, even allowing for greater body size.

Two other groups, both at different times included in the order Primates, are here considered as separate orders—the tree shrews (order Scandentia, pp144–149) and the flying lemurs (order Dermoptera, pp150–151)—which reflects the present scientific consensus.

The primates occupy a wide range of habitats and show a wide diversity of adaptations to their contrasting environments. The order contains terrestrial species as well as arboreal ones; species active at night as well as those active by day; specialized insectivores as well as fruit- and leaf-eaters. Primates range in weight from less than 3.5oz (100g) in the Dwarf bush baby, to over 220lb (100kg) in the gorilla, both inhabitants of the wet forests of tropical Africa. As a result, virtually all aspects of their biology vary widely from species to species (the following comments relate to primates other than humans). Gestation lengths range from around two months to eight or nine months; the weight of newborn infants from less than 0.4oz (10g) to over 4.4lb (2,000g). Weaning age differs from less than two months to over four years, and age of first breeding from nine

▶ **Largest and smallest** of the higher primates. Weighing in at some 350lb (160kg), the male African gorilla BELOW is a thousand times heavier than the Pygmy marmoset of South America ABOVE. Some prosimians are even smaller, such as the two lesser mouse lemurs and the Dwarf bush baby.

▼ **Posture and gait.** Some lemurs (**a**) walk on all fours and the tree-dwelling life-style of most species is reflected in the longer hindlimbs for leaping. Other prosimians, such as the tarsier (**b**), are "vertical clingers and leapers" adapted for leaping between vertical trunks. Most arboreal monkeys such as the Diana monkey (**c**), have well-developed hindlimbs and a long tail for balancing: baboons and other ground-dwelling monkeys have forelimbs as long as, or longer than, their hindlimbs. Among apes, the knuckle- walking gorilla (**d**) is the most terrestrial. The gibbons (**e**) arm-swing beneath the branches on their long arms and, like the chimpanzee (**f**), may walk upright, thus freeing the hands.

ORDER PRIMATES

Includes **Night monkey** (*Aotus trivirgatus*), **titis** (*Callicebus* species).

Marmosets and tamarins
Family Callitrichidae
Twenty-one species in 5 genera.
Includes **Common marmoset** (*Callithrix jacchus*), **Emperor tamarin** (*Saguinus imperator*).

Old World monkeys
Family Cercopithecidae
Eighty-two species in 14 genera.
Includes: **Hamadryas baboon** (*Papio hamadryas*), **gelada** (*Theropithecus gelada*), **mandrill** (*Papio sphinx*), **Savanna** or **"Common" baboon** (*P. cynocephalus*), **Barbary macaque** (*Macaca sylvanus*), **Japanese macaque** (*M. fuscata*), **Toque macaque** (*M. sinica*), **Guinea forest red colobus** (*Colobus badius*), **Guinea forest black colobus** (*C. polykomos*), **guenons** (*Cercopithecus* species), **Hanuman langur** (*Presbytis entellus*), **talapoin** (*Miopithecus talapoin*).

Great apes
Family Pongidae
Four species in 3 genera.
Gorilla (*Gorilla gorilla*), **Common chimpanzee** (*Pan troglodytes*), **Pygmy chimpanzee** or **bonobo** (*P. paniscus*), **orang-utan** (*Pongo pygmaeus*).

Lesser apes
Family Hylobatidae
Nine species of the genus *Hylobates*.
Includes **siamang** (*H. syndactylus*) and **gibbons.**

Man
Family Hominidae
One species, *Homo sapiens*.

[*] The entire order Primates is listed by CITES.

months to nine years. One exception to this diversity is litter size, which seldom exceeds two in any species, while the majority of primates produce a single offspring at a time.

The ecology of primates shows contrasts that are just as striking. In some lemurs, individuals may live their whole lives within 660ft (200m) or less of their birthplace while in species such as the Hamadryas baboon individuals regularly move over 9.3mi (15km) per day. Population densities vary between species from less than three animals per square mile to over three thousand.

On a wider geographical scale, primates are almost totally confined to the tropical latitudes between 25°N and 30°S (see map). Among the exceptions are two species of macaques: the Barbary macaque, sometimes misnamed the Barbary ape, found in North Africa and on the Rock of Gibraltar; and the Japanese macaque, which occurs in both the main islands of Japan. The restriction of primates to the Tropics and Subtropics probably stems from their dependence on diets consisting largely of fruit, shoots or insects—items which are scarce during winter in temperate regions.

An oft-quoted list of morphological characters common to all primates was drawn up in 1873 by the English biologist St George Mivart: primates were "unguiculate, claviculate, placental mammals, with orbits encircled by bone; three kinds of teeth, at least at one time of life; brain always with a posterior lobe and calcarine fissure; the innermost digits of at least one pair of extremities opposable, hallux with a flat nail or none; a well-developed caecum; penis pendulous; testes scrotal; always two pectoral mammae."

In practice, this is a very general blueprint. Furthermore, none of the traits it lists is on its own peculiar to primates: many other mammals have clavicles, three kinds of teeth and a pendulous penis. This absence of specialization reflects the diversity of lifestyles found within the order.

The different evolutionary lines of primates show a number of other common trends. One of the most important is the gradual refinement of hands and feet for grasping objects. This is associated with the development of flat nails on fingers and toes superseding the claws of ancestral primates, the evolution of sensitive pads for gripping on the underside of fingers and toes, and the increasing mobility of individual digits especially in the thumb and big toe culminating in the human hand, in which the thumb can be rotated to oppose the fingers, thus

THE BODY PLAN OF PRIMATES

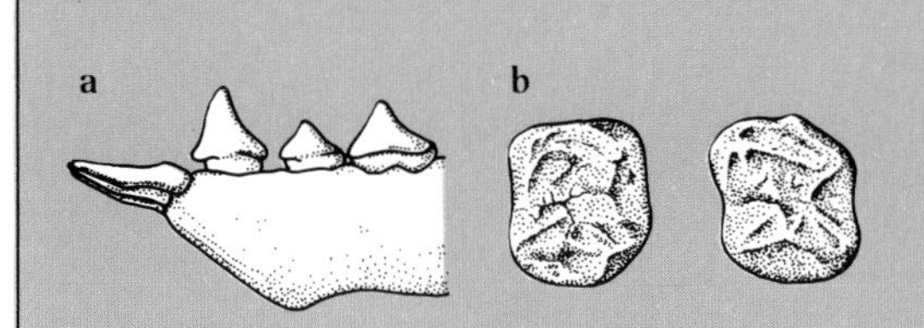

▲ **Teeth.** Insectivorous precursors of primates had numerous teeth with sharp cusps. In prosimians such as *Lemur* (**a**), the first lower premolar is almost canine-like in form, while the crowns of the lower incisors and canines lie flat to form a tooth-comb, as in bush babies, which is used in feeding and grooming. In leaf-eating monkeys of the Old World, such as *Presbytis* (**b**), the squared-off molars bear four cusps joined by transverse ridges on the large grinding surface that helps break up the fibrous diet. In apes such as the gorilla (**c**) the lower molars have five cusps and a more complicated pattern of ridges.

► **Hands** (left) **and feet.** The structure of primate hands and feet varies according to the ways of life of each species. (**a**) Hand of a spider monkey, showing the much reduced thumb of an arm-swinging species. (**b**) Gibbon: short opposable thumb well distant from arm-swinging (brachiating) grip of fingers. (**c**) Gorilla: thumb opposable to other digits, allows precision grip. (**d**) Macaque: short opposable thumb in hand adapted for walking with palm flat on ground. (**e**) Tamarin: long foot of branch-running species with claws on all digits except big toes for anchoring (all other monkeys and apes have flat nails on all digits). (**f**) Siamang and (**g**) orang-utan; broad foot with long grasping big toe for climbing. (**h**) Baboon: long slender foot of ground-living monkey.

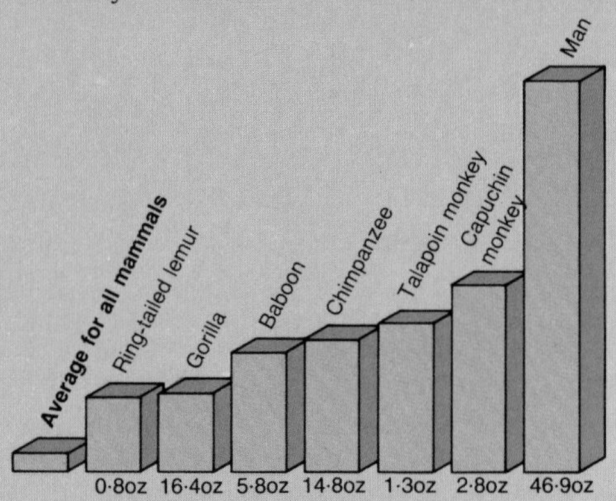

► **Relative brain size.** The degree of flexibility in the behavior of a species is related to both absolute and relative brain size. It is no surprise that, in terms of actual brain weight (figures shown in ounces), the great apes are closest to man. But when comparison is based on an index that allows for the influence of body size on the size of the brain, it is the versatile Capuchin monkey that turns out to be closest to man.

▼ **Skeletons.** The quadrupedal lemurs and most monkeys, like the guenons (**a**), retain the basic shape of early primates – long back, short, narrow rib-cage, long, narrow hip bones, and legs as long as or longer than the arms. Most live in trees and move about by running along or leaping between branches. Their long tail serves as a rudder or balancing aid while climbing and leaping. Ground-living monkeys, such as the baboons, generally have more rudimentary tails.

Neither apes nor the slower-moving prosimians have tails. In the orang-utan (**b**) and other apes, the back is shorter, the rib-cage broader and the pelvis bones more robust – features related to a vertical posture. Arms are longer than legs, considerably so in species, such as the gibbons and orang-utan, that move by arm-swinging (brachiation). Further dexterity of the hands has accompanied the development of the vertical posture in apes, some of which (as more rarely some monkeys) may at times move about bipedally like man.

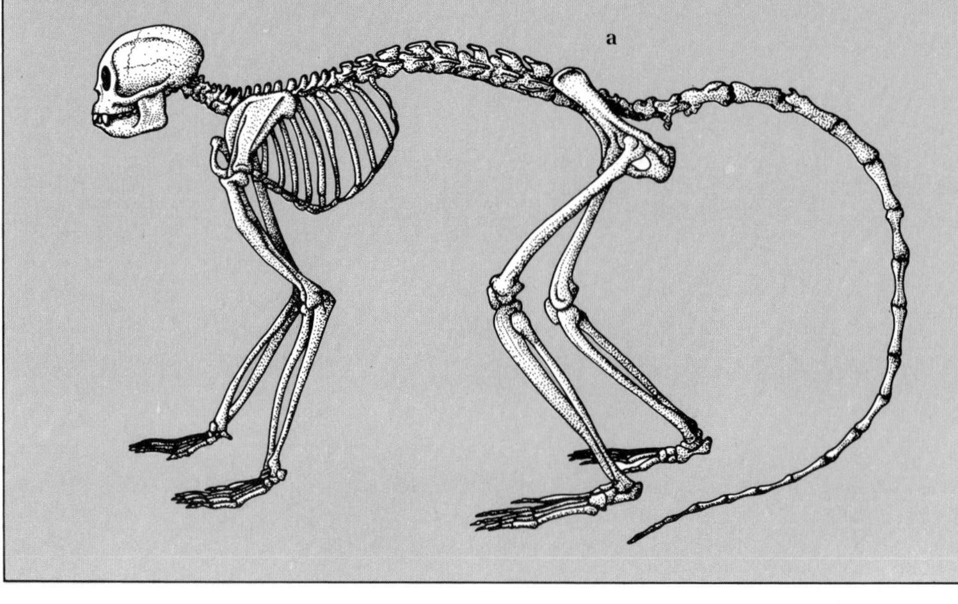

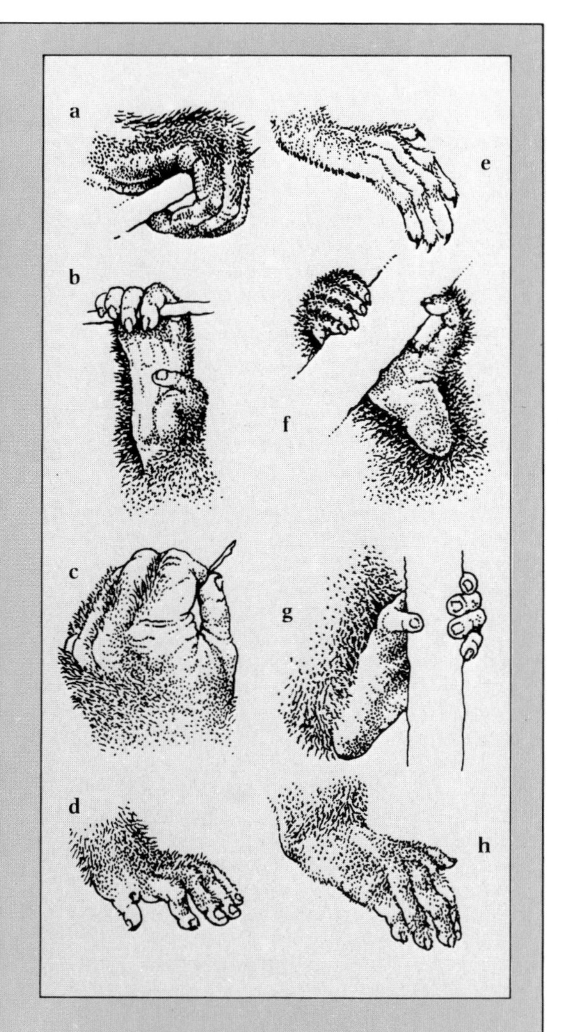

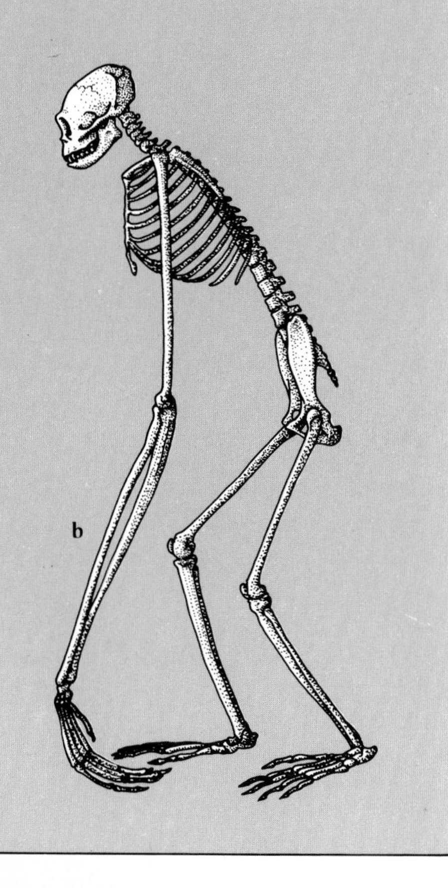

providing a grip that is both powerful and precise.

A second evolutionary trend is the gradual foreshortening of the muzzle and flattening of the face. This is associated with a general decline in the importance of smell and increased reliance on vision, leading eventually to efficient stereoscopic color vision.

There has also been a progressive increase in the relative and absolute size of the brain, with special elaboration and differentiation of the cerebral cortex and decline in the relative importance of the olfactory centers of the brain, associated with increased dependence on sight.

Other trends in primate evolution are a general reduction in the rate of reproduction, associated with protracted maternal care, delayed sexual maturity and extended life-spans, and a progressive dependence on diets of fruit and/or foliage, with a reduction in the proportion of animal matter eaten.

Finally the evolution of primates has brought an increasing complexity of social behavior with a progressive shift from a breeding system in which males and females occupy overlapping territories or home ranges towards a diverse array of breeding systems, including monogamous pairs, harem polygyny (where single males monopolize access to groups of females) and multi-male polygyny (where troops include breeding adult males).

Evolution

On an evolutionary time scale, the primates are a very recent phenomenon. Animals have existed on earth for over 600 million years and mammals for at least 200 million. In contrast, the first known primate appeared around 70 million years ago, the first hominids (*Australopithecus*) at least 6 million years ago, the first men (*Homo sapiens*) at least 300,000 years ago, and the modern subspecies (*H. s. sapiens*) no more than 50,000 years ago.

The first primates were probably superficially comparable to present-day tree shrews (see pp144–145)—arboreal, quadrupedal omnivores weighing around 5oz (150g) and obtaining their food on the ground and in the lower levels of tropical forests.

During the Paleocene and Eocene (65–35 million years ago), primates are known to have radiated throughout North America and Europe. They ranged in size up to that of the larger modern lemurs. The low-cusped cheek teeth of some early forms suggest that they were herbivores, while others were probably fruit-eaters or insectivores. Most were quadrupeds, but the skeletons of some resemble those of present-day tarsiers and they were probably active leapers, able to spring from stem to stem.

A large number of lemur- and tarsier-like species existed during this period, but the fossil record is patchy and it is impossible to identify the ancestors of present-day primates with any certainty. It was probably during the Eocene that the ancestors of the modern prosimians (lemurs, lorises and tarsiers) first became distinct from the evolutionary line that led to the monkeys and apes (anthropoids), subsequently radiating to form the main groups of present-day prosimians. Fossil relatives of both lesser apes (gibbons and siamang) and the greater apes (chimpanzees, orang-utan and gorilla) are known from the early Miocene, some 20 million years ago, and it is possible that these two groups may have diverged some time before this.

To judge from the scientific literature on primate evolution, one is easily misled into thinking that the most important differences between primates lie in the number and shape of their teeth. This is not because these represent the most important adaptive contrasts between groups (though in many cases changes in the tooth shape and size are closely related to the occupation of new ecological niches), but simply because teeth

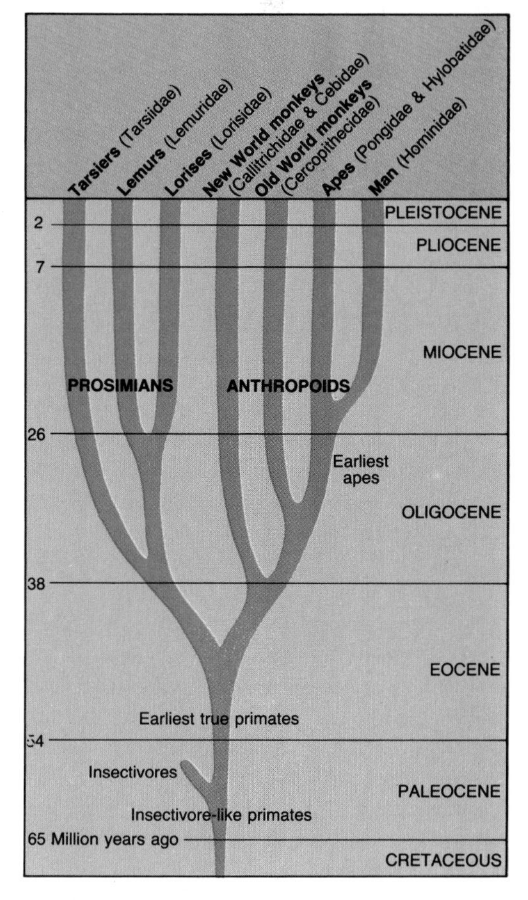

make better fossils than other parts of a primate's skeleton—with the result that much of our knowledge of evolutionary change is based on tooth structure. After teeth, skulls are the next most durable part, and then the heavier bones such as the femur and the pelvis.

As the early primates diverged progressively from their insectivorous ancestors, they required fewer teeth but a bigger tooth surface area for grinding the fruit and vegetation that came to form a larger part of their diets. These ancestors probably had at least 44 teeth—eleven on each side of their upper and lower jaws. In the early primates, tooth number fell rapidly and, today, it varies from 18 to 36. Dental formulae differ consistently between primate families and offer one of the easiest ways of identifying specimens.

Many other dental characteristics are important in distinguishing between primate families. In most of the lemurs and all the bush babies, pottos and lorises, the upper incisors are reduced in size and, in the lower jaw, they and the lower canines lie flat (procumbent), forming a comb that is used in feeding and grooming. The premolars of many primitive mammals rise to a single point (cusp), whereas in the monkeys and apes they mostly have two or more cusps. The Old World monkeys have four main cusps on their molars with transverse ridges running from side to side between them—an arrangement thought to improve their ability to eat coarse fruit and vegetation—whereas apes and men have five principal cusps on their lower molars and a more complex arrangement of connecting ridges. Finally, the apes (and Old World monkeys) differ from men in the greater development of their canine teeth as well as in the shape of their first lower premolar, which in apes and monkeys (but not in men) is modified to provide a shearing edge down which runs the sharp blade of the upper canine. This difference between men and the other higher primates (monkeys and apes) may be associated with contrasting use of weapons against competing members of the same species or against predators—the monkeys and apes had to rely on canines as the ultimate deterrent, whereas men had more effective weapons.

As the primates developed, they adopted ways of life that required them to process more and more information. As a result, brain size grew—producing a range today from around 0.122cu in in the smallest prosimians to over 76.25cu in in modern man—in all cases representing an increase in the size of the brain relative to that of the body. This may have been closely associated with changes in ranging behavior for, among present-day primates, most species that typically live in large home ranges have comparatively big brains for their body size. Changes in brain size were also associated with a gradual change in the shape of the skull, but its basic structure varied little. However, the pattern of joining between the smaller bones on the side of the skull changed, and can be used to distinguish present-day primate groups.

As primates radiated into different habitats, the shape of their bodies became adapted to suit their environments. The earliest primates mostly walked on all four legs, like many of the present-day lemurs—their rib cages were narrow, their backs long and the blades of their pelvises long and narrow. Today's quadrupedal lemurs and monkeys, such as the Ring-tailed lemur and the baboons, retain this basic shape, but those that adopted a vertical posture (either because they usually moved by leaping from vertical stem to vertical stem, or because they swung by their arms) like the tarsiers, sifakas, spider monkeys and apes, became progressively barrel-chested and their backbones and pelvises shortened and became more robust.

The relative lengths of arms and legs also

▼ **Feeding ecology** Many primate species may share the same forest by splitting up their environment and so reducing competition for food. In this "share out" by means of natural selection, differences in feeding times, kinds of diet and levels where the animals forage are all important.

In the West African rain forests, five different species of lorisids commonly forage by night: the diminutive Dwarf bush baby feeds mainly on insects in the upper levels, which are used also by the fruit-eating potto. The fruit-eating Needle-clawed bush baby *Galago elegantulus* uses the middle and lower levels, while Allen's bush baby is mostly confined to forest floor shrubs, as is the insect-eating angwantibo.

Among the day-active species in the same forest is the omnivorous mandrill, which obtains most of its food from the ground or shrub layer, as does the leaf-eating gorilla. The chimpanzee, a fruit-eater, uses all levels of the forest. Also in the upper levels, the Red colobus monkey has to feed on many species for its diet of leaf shoots, flower buds and flowers, whereas the Guinea forest black colobus (or Black-and-white colobus) also eats mature leaves and ranges less widely.

▲ **A feast of flowers:** the Squirrel monkey of Central and South America feeds chiefly on fruit and insects and may supplement this diet with the occasional tree frog or even flowers. This smallest of cebid monkeys also lives in some of the largest troops (30–40 animals) and it is found to have a relatively large brain when the effect of body size is taken into account.

▶ **Leaf-eating giants** OVERLEAF. The gorilla, the largest of all living primates, feeds extensively on leaf material and the social groups travel relatively little during the day. Their close relatives, the chimpanzees, by contrast, feed mainly on fruit and they typically split up into very small groups that travel quite widely in search of food each day.

changed as different primates came to use different forms of locomotion. Quadrupedal primates have arms and legs of similar length, whereas the vertical-clingers-and-leapers (such as the tarsiers) have much more developed hind limbs, providing the power for their long springs. In contrast, in the arm-swinging (brachiating) apes, the relative size of the arms has increased while leg length has reduced. Finally, in man, leg length has again increased, the arms are shorter, and the shoulder blade (scapula) is rotated backwards on the rib cage.

The contrasting locomotor behavior is also reflected in the shape of hands and feet. Quadrupedal species retain a small, slightly divergent thumb. In brachiating monkeys, the thumb is greatly reduced or even missing altogether—presumably because it snagged on the branches which they gripped, and impeded movement. In contrast, in the apes, the thumb is well developed and mobile, providing a precise and powerful grip when opposed to the rest of the fingers.

The feet show related changes. In the quadrupedal lemurs and monkeys, the feet are long and narrow, whereas in the brachiating great apes they are broader with big toes adapted for more powerful grasping. In men, the bones (phalanges and metatarsals) in all the toes are reduced and are straighter and heavier, the big toe is the longest, and the weight is borne on the ball of the foot and the heel.

Finally, the structure and length of the tail varies widely. Most primitive primates probably had substantial tails like many of the present-day lemurs. These may have helped the animals to balance as they maneuvered along slender branches. Among prosimians, tails are retained by most of the smaller active leapers but lost in the slow-moving potto and lorises. Tails are present in most of the monkeys but show an important contrast: prehensile (gripping) tails often with a special sensitive pad at the tip, have only developed in some South and Central American monkeys. Tails have been lost in some terrestrial Old World monkeys, in all the present-day apes and in men.

Primate Ecology

In order to survive and breed successfully, all animals must obtain adequate food. Even where food supplies appear to be superabundant, as in many of the tropical rain forests, particular components of the diet are often in short supply and competition for these can be intense. This does not necessarily mean that individuals fight over particular food items—some may simply collect food more quickly or efficiently than their neighbors.

Competition for food is likely to favor the evolution of contrasting feeding behavior in different species which share the same environment. Different animals split up their environment in different ways—for example, many invertebrates feed on one part of a particular plant species. Primates are more eclectic in their food choice and most species will eat a wide array of foods. However, species occupying the same habitat differ considerably in the timing of feeding activity, in the levels of the forest from which they obtain food, in the type of food eaten (insects and other animals, gums, fruit, flowers, or foliage) and in the range of different foods that their diet encompasses.

Even when two species apparently eat the same food at the same levels, close inspection normally shows that they differ in some important aspect of their food choice. For example, the Guinea forest red and Guinea forest black colobus monkeys are both leaf-eaters (folivores) which use the upper levels of many of the same African rain forests, but they specialize in eating leaves at different stages of growth. Similarly, while chimpanzees and guenons are fruit-eaters (frugivores), the chimpanzees take mostly ripe fruit, while the monkeys also eat fruits at earlier stages of their growth.

Differences in feeding affect virtually all aspects of morphology, physiology and behavior. For example, all insectivorous primates are small in size, and the more folivorous species tend to be larger than their frugivorous relatives—thus the folivorous siamang is larger than the frugivorous gibbon and the folivorous gorilla is bigger than the frugivorous chimpanzee.

Species using similar habitats and foods often resemble each other in the structure of the teeth, the proportions of the body, the form and complexity of their digestive tracts, and the relative size of their brains, as well as many aspects of social behavior and ecology. Folivorous primates are usually more sedentary than frugivorous ones—they require smaller home ranges and their population densities are higher. For example, in the Madagascan forests, troops of the omnivorous Ring-tailed lemur have home ranges of around 17.3 acres (7 hectares), whereas those of the leaf-eating Brown lemur seldom exceed 2.4 acres (1 hectare). The extent of a species' home range is also related to body size (bigger primates require more food and have larger home ranges than smaller ones) and also to the nature of the habitat.

The interaction between the primates and the forest is thus a complex one, and easily disrupted by human interference, for example, hunting for food. The collection of primates for use in medical research has made damaging inroads into some populations: during the 1950s, around 200,000 macaques were imported into the USA alone from Asia each year, and even today the USA still imports around 20,000 wild-caught primates each year. But it is habitat destruction, rather than trapping or hunting, that has had the most important influence on primate populations.

The world's rain forests are being felled at an alarming rate to provide timber or to make way for agriculture. In 1981 it was estimated that tropical rain forest was being

destroyed at a rate nearing 2.5 acres (1 hectare) a second—that is, an area the size of France every two years, or of the State of Colorado every year. Where forests are replanted, exotic conifers are often used which provide no food for primates, or eucalyptus, to which the animals have to become accustomed. At the moment, the populations of forest primates are dwindling fast and many species will soon only be found in isolated reserves or will have disappeared altogether. Of the 22 species of Madagascan lemurs, over half are in danger of extinction or are likely to become threatened in the near future. Twenty-one species of monkeys in Central and South America, 14 species of Old World monkeys, 4 gibbons and all the 4 great apes are also endangered or threatened. Some species—such as the beautiful Golden lion tamarin—are now so rare in the wild that captive breeding colonies have been established to ensure that they are not lost (see p55).

Primate Societies

One of the most striking characteristics of the primates is their sociality. With few exceptions, monkeys and apes live in groups and so, too, do the majority of the prosimians. Moreover, unlike many other social animals in which groups vary in membership from hour to hour as individuals come and go, primate groups are largely stable in their membership.

It is not obvious why primates live in groups, for feeding in groups inevitably increases competition for food items. It is notable that while virtually all species active in daytime live in groups, most nocturnal species are largely solitary. One exception is the Night monkey of the South American forests (see p68), which lives in small groups of 2–5 animals, probably centered on a monogamous breeding pair. In some bush babies, related females share the same nest during the daytime but, with the exception of mothers and dependent offspring, the nocturnal bush babies usually forage alone.

Perhaps the most likely reason why nocturnal primates are solitary is that most rely on hiding rather than rapid flight to escape from predators—and, clearly, an individual can hide more effectively than a group. Solitary behavior may also be encouraged by greater success in hunting arthropods, and by the need for a larger home range (relative to body weight) for insectivores as opposed to leaf- and fruit-eating primates.

The largest primate groups occur among the terrestrial baboons and macaques. For example, Gelada baboons, which are found in the arid highlands of Ethiopia, sometimes collect in herds of up to 1,000 animals, and most baboons and macaques live in groups of between 20 and 50 animals. Ground-living primates may be more vulnerable to predators than arboreal species. Living in big groups may increase the probability that predators will be detected, and by placing itself close to other animals, the individual reduces the chance that it will be the victim.

Among primates active in the daytime, the smallest social groups are found in the apes. Gibbons are usually seen in groups of 3–4, while both orang-utans and chimpanzees spend much of their time feeding alone. One possible explanation is that feeding competition is especially intense in these species since they depend principally on ripe fruit and their large body size is associated with large nutritional requirements.

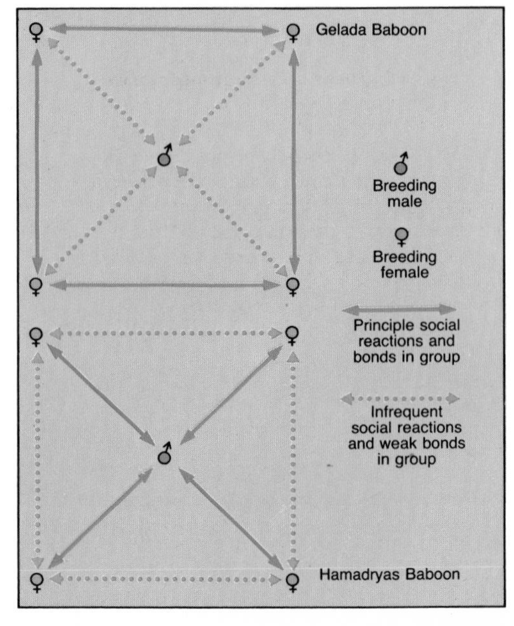

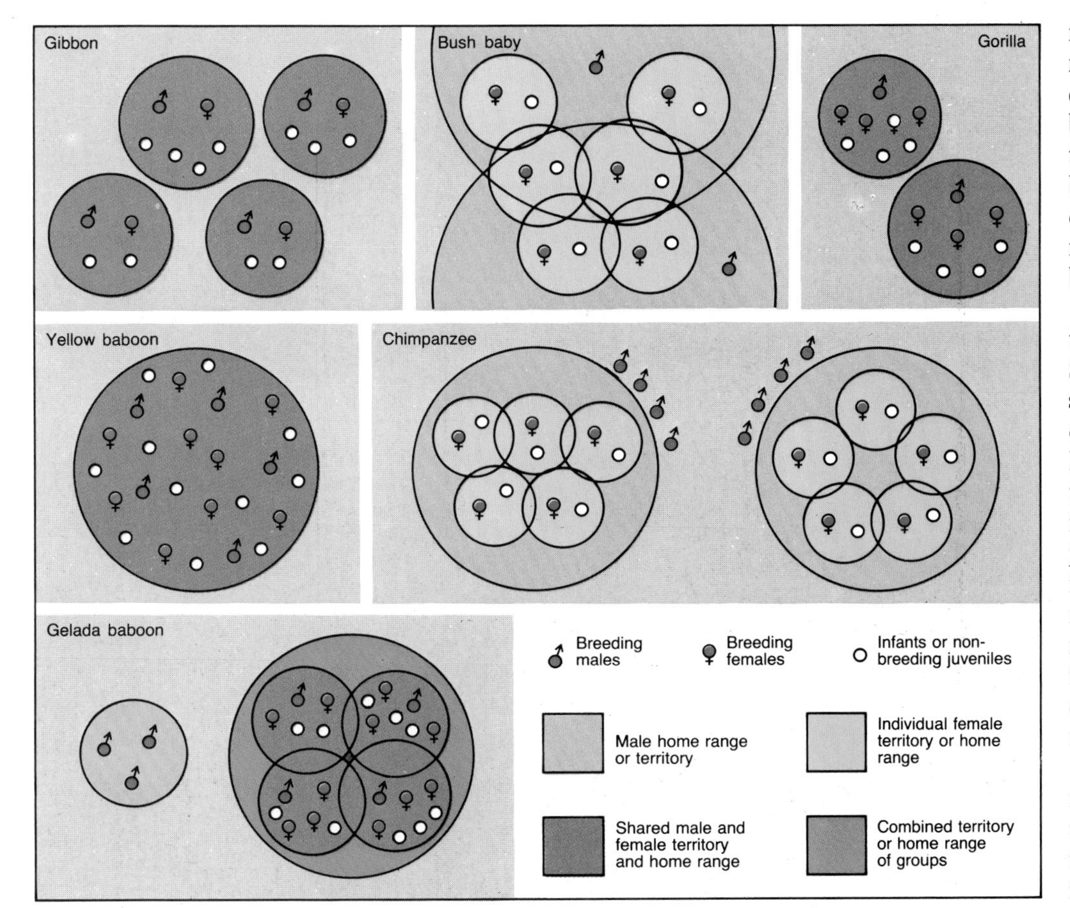

▲ **Types of primate society.** Most primates live in groups. Gibbons live in monogamous pairs, up to four immature offspring remaining with their parents. In bush babies the ranges of several related females and their offspring overlap; males occupy larger areas that include the ranges of several females. Gorilla groups comprise a single breeding male with a harem of females together with their young. Yellow (or Common) baboons live in multi-male troops with several breeding males and many breeding females, while harem groups of Gelada baboons aggregate into larger bands which exclude nonbreeding mature males. In chimpanzees, communities of unrelated females with individual home ranges, but considerable overlap, are monopolized by groups of related breeding males.

◄ **Social interactions.** Like many other Old World monkeys, both Gelada and Hamadryas baboons usually live in harem groups. In the Gelada baboon (photograph, ABOVE) the principal social bonds are between females (indicated by solid lines in the diagram); the breeding male mates with the females but otherwise interacts with them relatively little (broken lines). In the Hamadryas baboon it is the dominant male that maintains cohesion of the group, by continually herding his females together, and social bonds between females are relatively weak.

Comparing primate societies in terms of group size alone is misleading, for it masks the fact that groups differ in structure. Even among the solitary species "social" behavior often differs between the sexes: males usually defend extensive territories which overlap the smaller home ranges of several females. Subordinate males are excluded from access to females and adopt home ranges of their own, often in less favorable habitat.

For many species that are usually seen in small groups these appear to consist of a monogamous breeding pair and their offspring. These species include lemurs (Lemuridae) and the indri of Madagascar, the marmosets, titis and Night monkey among the New World monkeys and the gibbons among the apes. On reaching maturity, the young leave the group to seek mates.

Among the Old World monkeys, many species are usually found in harem groups, consisting of a single breeding male and a number of mature females. These species include most guenons, the Guinea forest black colobus, several species of langurs, the Hamadryas and Gelada baboons and, among the apes, the gorilla.

In some of these species, young or subordinate males are allowed to remain in the troop but seldom get the opportunity to mate, while in others, including the Hanuman langur, sexually mature males are excluded from the group and form all-male bands (see p114). In most of these species, the males of neighboring troops are hostile, but in the Hamadryas and Gelada baboons of open terrain and in the forest-dwelling mandrill, harem units aggregate in larger bands or herds.

The howler monkeys of South and Central America, the red colobus monkeys and Savanna baboons of Africa, among other species, usually live in multi-male troops, consisting of several breeding males and a larger number of females. It was once thought that the presence of several males in a troop served to protect females and juveniles from attack by predators, but there is little evidence that this is the case. The most likely explanation for the occurrence of multi-male troops is that they occur where female groups are too large or too widely dispersed for it to be feasible for a dominant male to evict other males.

Finally, chimpanzees, and the spider monkeys of the New World, are found in unstable groups which vary in size from day to day, but belong to stable larger communities. In chimpanzees, females appear to have individual home ranges which overlap widely and they will collect in groups at particularly rich feeding sites. Bands of related males monopolize breeding access to communities of females, maintaining hostile relations with the males of neighboring communities.

In some species that live in harem groups, the social bonds that maintain the group run principally between the females, while in others they exist principally between the harem male and his females. In most of the monogamous primates, both male and female offspring disperse from their natal group after adolescence, while in the majority of species living in harems or multi-male groups, males disperse after adolescence and young females remain in the group, often indefinitely. As a result, females belonging to the same troop are usually related to each other and the social bonds between them are strong, whereas males belonging to the same troop are unlikely to be closely related. However, in some species, including the Hamadryas baboon, the red colobuses, and the Common chimpanzee, females disperse while males commonly remain in their natal band. In these species adult females belonging to the same troop are rarely closely related and social bonds between them are usually weak. Males in the same troop or community are often closely related and commonly assist each other in excluding unrelated male would-be immigrants.

The nature of the breeding system is also associated with the extent to which the sexes differ in other characteristics. In strongly polygynous species, males compete intensely to gain breeding access to females and their fighting ability is often important. Males of these species are typically much larger than females, whereas in virtually all monogamous primates males and females are the same size. In addition, in polygynous species, male weaponry is more developed: males have relatively larger canines than in monogamous species and often have heavy ruffs or manes. In female monkeys and apes, pronounced sexual swellings are found principally in species where females have access to several different mating partners—they occur in the baboons, red colobus monkeys and talapoin, all of which live in multi-male troops—whereas in species living in harems (with the exception of the Hamadryas baboon) such swellings are not found.

In summary, monogamous societies appear to occur in species whose food requirements are concentrated in areas small enough to be defended effectively. Where this is not the case, females usually form groups unless feeding competition is so intense that it outweighs the advantages of grouping. Where these groups are too large to be defended effectively by a single male, several males join the group and large troops consisting of members of both sexes are formed.

Social Relationships

Social structure is ultimately a consequence of the action of individuals. An important breakthrough in primate research was the appreciation that many aspects of social behavior represented adaptations towards maximizing the individual's reproductive success or the reproductive success of its offspring.

The infant primate is dependent on its mother for much of its food. This period of dependence varies from around two months in the smallest species to over a year in the larger ones. As well as providing milk, the mother protects the infant from predators and from competition with older members of the same group. Experimental studies have shown that maternal care and the quality of the mother–infant bond can also have a profound effect on the subsequent development of the infant's social behavior and on its social relationships as an adult.

The mother is not the only adult that helps to care for the infant. In many monogamous primates, including the siamang, Night monkey and marmosets, the male also carries and looks after the infant (see titi monkeys, p70, for instance). Since there is a high probability that they have fathered them, males are contributing to their own breeding success by doing so.

In the marmosets and tamarins, older siblings carry and care for their younger brothers and sisters and, in several other primate species, adult females belonging to the same social group as the mother show an intense interest in her infant and will groom and carry it. This behavior, misleadingly known as "aunting" (for it is not confined to aunts), appears to be commonest where social groups consist of closely related females.

Relationships between infants and other adults are not always tolerant or friendly. In some species that live in harem groups members of all-male bands outside the group regularly attempt to displace the harem male and gain a harem of their own. When they succeed, they commonly attempt to kill infants fathered by the previous male.

Even related females sometimes attack and can kill infants. In Toque macaques, female juveniles and subadults are more frequently threatened and displaced from feeding sites than males of the same age and are substantially more likely to die from starvation. One reason why adult Toque macaques may be more intolerant of young females than of young males is that most females remain in the natal group whereas males disperse after adolescence. To an adult female, therefore, a juvenile female may represent a future competitor both to herself and to her offspring.

Within primate groups, adult males and females are usually ordered in a dominance hierarchy. An individual's rank is reflected in its priority of access to food, sleeping sites and mates and, in both sexes, rank can have an important influence on breeding success. The factors determining rank are often hard to identify, but in many of the baboons and macaques kinship is evidently important. In these species females associate closely with other members of the same matriline within their troop and support them in contests with members of other kin groups. A female's rank depends on the matriline to which she belongs: all members of high-ranking matrilines out-rank all members of low-ranking ones, irrespective of age. Dominance relationships between females belonging to the same matriline are also consistent and, in baboons and macaques,

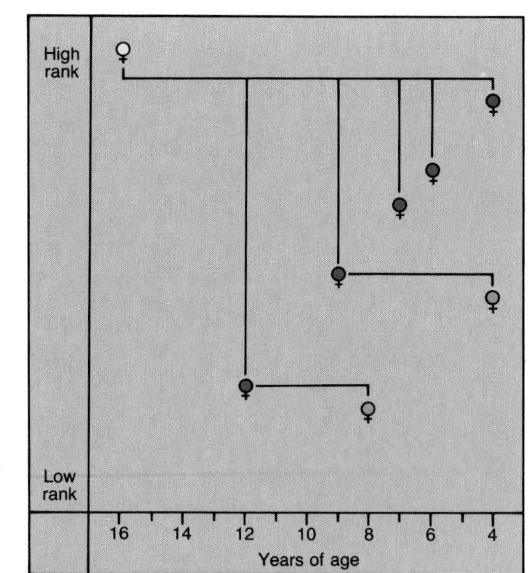

▲ **Rank and birth order.** Dominance hierarchies of males and females are found in most primate groups. Among many baboons and macaques males leave the group after adolescence, while females remain. Social bonds between these related females maintain the cohesion of the troop, which is therefore matrilinear. The rank of different members of the same matriline depends on birth order: daughters rank immediately below their mother and *above* older sisters. In due course a daughter may mate and found her own matriline. All members of a matriline share the same ranking in relation to members of other matrilines.

► **Mutual grooming among Rhesus macaques.** Such behavior is prominent in interactions between group members in many social primate species.

► **Facial expression** is more varied in higher primates than other animals, and in the chimpanzee particularly. (1) Play face: relaxed, open mouth, upper teeth covered. (2) Pout: used in begging for food. (3) Display face: used in attack (see p130) or otherwise to show aggression; facial hairs erected. (4) Full open grin: intense fear or other excitement. (5) Horizontal pout: shows submission, eg when whimpering after being attacked. (6) Fear grin: during approach to or from a higher-ranking chimpanzee.

have proved to be inversely related to priority in order of birth.

Closely related adult females often cooperate in other ways. They will share feeding sites and sleeping sites and care for each other's offspring. However, cooperative relationships are not restricted to relatives. In Olive baboons, males that are attempting to displace a consorting male and take over a receptive female will enlist support from one or more others and will, in their turn, provide assistance in future to their helpers. The males benefit by reciprocating, for their breeding success can be greatly enhanced.

The complex network of social relationships within primate troops requires an elaborate and accurate system of communication. Among the nocturnal lemurs and lorises, olfactory communication usually plays a major role and individuals mark their territories with urine, feces, or the secretion of specialized glands, which can signal the animal's sex, reproductive status, and individual identity. However, a system of communication based on smell is both too slow and too limited to meet the needs of the gregarious primates, which have evolved elaborate visual and vocal signalling systems, based on continuously varying signals which can convey subtle nuances of meaning; these systems have culminated in human language.

However, the gap between the communication systems of non-human primates and human language is a big one, for though non-human primates can express their own emotions and intentions, they rarely, if ever, communicate about their environments. Attempts to teach chimpanzees human language (either speech or Ameslan, the American sign language devised for use by the deaf) have shown that they possess more complex abilities than they appear to use in the wild. However, though they can learn to associate vocal or visual signals with particular objects and even to give primitive instructions, their ability to structure sentences or give complex instructions is much more limited than that of human children. THC-B

PROSIMIANS

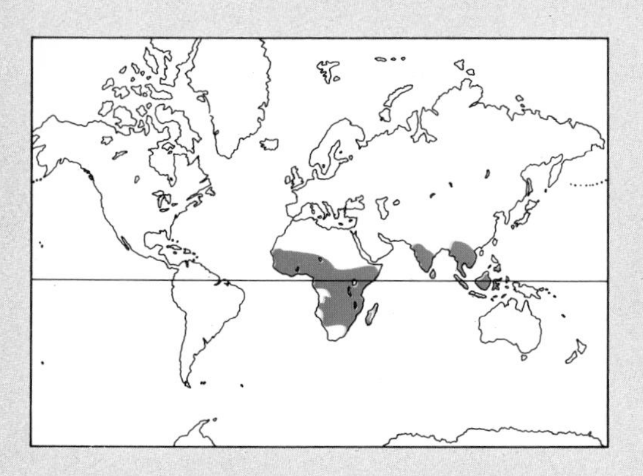

Suborder: Prosimii
Thirty-five species in 18 genera and 6 families.
Distribution: Madagascar, Africa, S of Sahara, S India, Burma to Indonesia and Philippines.

Habitat: chiefly forests.

Size: from head-body length 4.9in (12.5cm). tail length 6.1in (15.5cm) and weight 1.9oz (55g) in the Brown lesser mouse lemur to head-body length 24in (60cm), tail 2in (5cm), weight 15–20lb (7–10kg) in the indri.

Lemurs (family Lemuridae)
Ten species in 4 genera.

Dwarf and mouse lemurs (family Cheirogaleidae)
Seven species in 4 genera.

Indri and sifakas (family Indriidae)
Four species in 3 genera.

Aye-aye (family Daubentoniidae)
One species, *Daubentonia madagascariensis.*

Bush babies, pottos and lorises (family Lorisidae)
Ten species in 5 genera.

Tarsiers (family Tarsiidae)
Three species of the genus *Tarsius.*

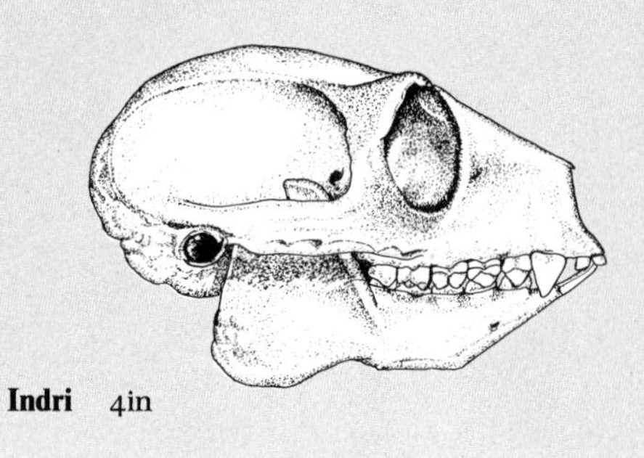

Tarsier 2in

Indri 4in

Many prosimians retain primitive characteristics and in this sense are closer to the ancestral primates than are the higher primates or simians (monkeys and apes)—hence the name given to the suborder into which six families are grouped: "forerunners of the simians." Of course, the present-day lemurs are not identical to the Eocene ancestors of monkeys, but their general structure and the range of ecological niches that they currently occupy may not be very different from those of Eocene lemurs (see p26). The largest group of modern prosimians are the lemurs, now usually grouped in four families all of which are today confined to the island of Madagascar lying off the coast of southeast Africa. How they arrived on Madagascar is still a mystery, for the island has apparently been separated from the African mainland since the Cretaceous, but this has sheltered the lemurs from competition with later primate families, including the monkeys and apes, and allowed them to radiate to fill a wide range of niches.

The dwarf lemurs (family Cheirogaleidae) are mostly small, in size ranging from the two lesser mouse lemurs, with an adult weight of 1.9oz (55g) or so, to the Greater dwarf lemur, which is as big as a Red squirrel. They are all nocturnal, arboreal and either eat fruit or insects, occupying a niche similar to that of the bush babies on the African mainland.

The family Lemuridae includes a wide range of animals varying in weight from around 1.1 to 6.6lb (500g–3kg). They include omnivores active by day (eg the Ring-tailed lemur), leaf-eaters active by day (eg the Brown lemur) and leaf-eaters active by night (eg the Sportive lemur). The majority of species are exclusively arboreal but some, like the Ring-tailed lemur, obtain much of their food from the ground. The indri and sifakas (family Indriidae), sometimes called leaping lemurs, differ from the true lemurs in that they are all vertical leapers and most are somewhat larger. They include one nocturnal species, the Woolly lemur, about which little is known, the fruit-eating indri active by day, and two species of sifaka, both of them leaf-eaters active by day. Little is known about the social life of the Woolly lemur, but the indri lives in monogamous pairs which defend their territories with elaborate calls, while the sifakas live in troops of 5–15 animals, usually including several adult males.

The last family of Madagascan lemurs consists of a single species, the aye-aye, a nocturnal insectivore without parallel in any other primate group. It is covered with coarse dark hair and the particularly thin

► **Huge eyes of the Philippine tarsier** indicate the nocturnal activity of these small prosimian predators from the islands of Southeast Asia.

▼ **The Thick-tailed bush baby** moves rapidly through the trees. It flees danger by running along branches, although it is a less agile leaper than the other, smaller bush babies.

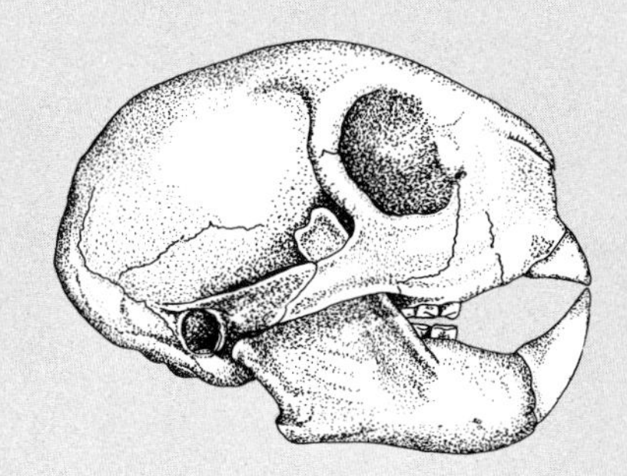

Aye-aye 3.4in

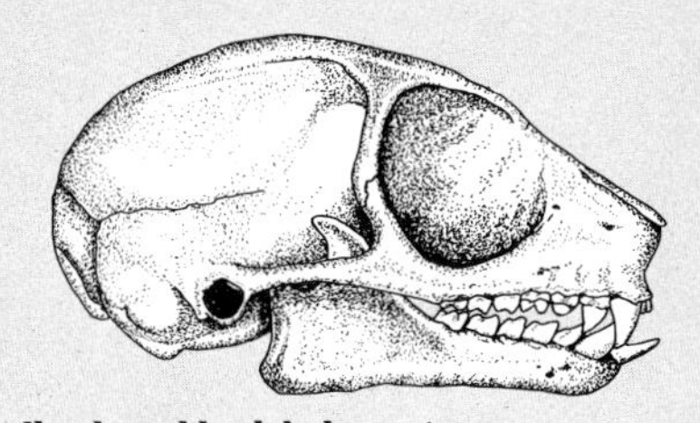

Needle-clawed bush baby 2in

Skulls of Prosimians

Prosimians have moderately developed jaws, relatively large eye sockets and a medium-sized, rounded brain-case. There is a complete bony bar on the outer margin of the eye socket (orbit). This bar provides support for the outer side of the eye. As in primates generally, the orbits are directed forwards for binocular vision. Orbits are relatively larger in nocturnal prosimians than in day-active species (some of the larger Madagascar lemurs) and the tarsier has by far the largest relative orbit size. The tarsier has a post-orbital plate of bone that confines the eye within an almost complete socket. It shares this feature with the higher primates (monkeys, apes and man).

Prosimians have relatively simple jaws and teeth compared to many mammals. The two halves of the lower jaw remain unfused throughout life. The molars are squared off in both upper and lower jaws and typically have four cusps, though tarsiers still have primitive three-cusped molars in the upper jaw. The molar cusps are generally low, but sharper, higher cusps are found in insect- or leaf-eating species. The common full dental formula of I2/2, C1/1, P3/3, M3/3 = 36 is found in dwarf lemurs (Cheirogaleidae), most medium-sized lemurs (Lemuridae) and all bush babies and lorises (Lorisidae), all of which have a tooth-comb in the lower jaw. The Sportive lemur (*Lepilemur*) differs from this basic formula in lacking the upper incisors, while members of the mostly day-active indri group (Indriidae) have lost the canines from the tooth-comb and have lost a premolar from each side of both jaws. The aye-aye (I1/1, C0/0, P1/0, M3/3 = 18) only retains its powerful, continuously growing incisors and a set of relatively tiny cheek teeth. Finally, the tarsiers have a unique dental formula of I2/1, C1/1, P3/3, M3/3 = 34 and no tooth-comb.

middle finger (which like the other fingers is elongated) bears a long, wiry claw on the tip which is used to extract the larvae of wood-boring insects.

The lemurs previously included a wide range of other "subfossil" species that have become extinct since the end of the Pleistocene. Some of these occupied ground-dwelling niches filled by hoofed mammals on the African mainland and one was as large as a pig (see box, p26).

Unlike the lemurs, the lorises and bush babies of the family Lorisidae are found across Africa and India to Southeast Asia and occur in the same habitats as many higher primates. They range in size from the diminutive Dwarf bush baby weighing about 2.1oz (60g), to the potto weighing over 2.2lb (1kg). Unlike the lemurs, they are all nocturnal, arboreal and either fruit-eaters or insectivorous: most of the forest areas where they are found are occupied by day-active monkeys and apes.

Found also in Southeast Asia, but confined to the offshore islands, are the three species of the tarsier family. Weighing slightly over 3.5oz (100g), tarsiers are nocturnal and insectivorous and live in low forest and undergrowth. Their most distinctive feature is the enormous development of their visual system and their huge eyes (in one species, a single eye weighs more than the brain). Present-day tarsiers are the relics of a much larger radiation of tarsier-like species during the Eocene. Tarsiers are superficially most like the bush babies. Their breeding system is much like that of the smaller nocturnal lemurs and lorises: they are solitary and males defend territories overlapping the ranges of females. THC-B

LEMURS

Twenty-two species in 12 genera.
Four families: Lemuridae, Cheirogaleidae, Indriidae, Daubentoniidae.
Distribution: Madagascar.

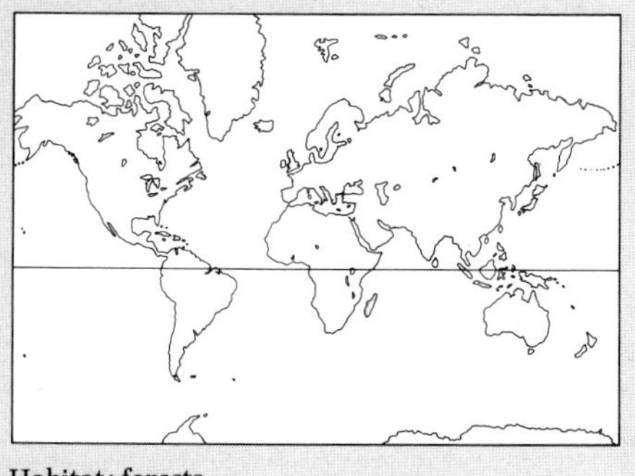

Habitat: forests.

Size: smallest are the lesser mouse lemurs—head-body length 4in (12.5cm), tail length 5.3–6.1in (13.5–15.5cm), weight 1.9–2.3oz (55–65g), and largest is the indri—head-body length 22.5–27.5in (57–70cm), tail length 2in (5cm), weight 15.5–22lb (7–10kg). The Hairy-eared dwarf lemur may weigh 1.6oz (45g).

Coat: soft in most species, often vividly colored.

Gestation: 60–160 days where known.

Longevity: not known in wild.

Lemurs (family Lemuridae, 10 species in 4 genera).
Typical lemurs (subfamily Lemurinae), 7 species including the **Ring-tailed lemur** (*Lemur catta*); **Black lemur** (*L. macaco*); **Crowned lemur** (*L. coronatus*); **Brown lemur** (*L. fulvus*) and subspecies including Collared lemur (*L. f. collaris*), Mayotte lemur (*L. f. mayottensis*) and Red-fronted lemur (*L. f. rufus*); **Mongoose lemur** (*L. mongoz*); and **Ruffed lemur** (*Varecia variegata*).
Sportive lemur (subfamily Lepilemurinae), 1 species, *Lepilemur mustelinus*.
Gentle lemurs (subfamily Hapalemurinae), 2 species, *Hapalemur griseus* and *H. simus*.

Dwarf lemurs (family Cheirogaleidae), 7 species in 4 genera including the **Gray lesser mouse lemur** (*Microcebus murinus*), **Brown lesser mouse lemur** (*M. rufus*) and **Coquerel's mouse lemur** (*M. coquereli*); the **Hairy-eared dwarf lemur** (*Allocebus trichotis*); the **Fat-tailed dwarf lemur** (*Cheirogaleus medius*) and **Greater dwarf lemur** (*C. major*); and the **Fork-crowned dwarf lemur** (*Phaner furcifer*).

Indri and sifakas (family Indriidae), 4 species in 3 genera including the **indri** (*Indri indri*), **Woolly lemur** (*Avahi laniger*), and the **sifakas** (2 species of *Propithecus*).

Aye-aye [E] (family Daubentoniidae), 1 species, *Daubentonia madagascariensis*.

[E] Endangered. Status of all lemurs is under review by IUCN.

Lemurs (the name means "ghosts") are confined to the island of Madagascar and represent the modern survivors of a spectacular adaptive radiation of primates that seems to have taken place essentially within Madagascar. The island, in effect, houses the results of a gigantic "natural experiment" in which ancestral lemurs were isolated at least 50 million years ago and gradually diversified into the modern array of over 40 species (including several large-bodied species which are sadly documented only by their subfossil remains). As such, lemurs have retained numerous primitive characteristics while at the same time developing many features in parallel to the monkeys and apes of the major southern land masses, especially among the larger day-active (diurnal) species. There is a trend in increasing body weight from the dwarf lemurs (Cheirogaleidae) through the medium-sized (Lemuridae) and on to the largest extant species in the family Indriidae. This trend is correlated with a shift away from predominant night-time (nocturnal) activity to exclusive diurnal activity, which broadly reflects the evolutionary trend among primates generally.

Lemurs

Like all the primates of Madagascar, the lemurs of the family Lemuridae evolved in isolation from the monkeys and apes of Africa, and from most of the competitors and predators facing primates on the mainland. Some play ecological roles similar to those of monkeys in Africa, while others have evolved in ways unique among the primates. About 2,000 years ago, people arrived on the island. They hunted, modified the habitat and introduced new species, particularly cattle and goats. Their evolution in isolation left the Malagasy primates ill-equipped to confront these new arrivals, and at least 14 species went extinct, including one whole subfamily of lemurs, the Megaladapinae, animals the size of orangutans that lived in the trees, and moved about like giant Koala bears. Today, three subfamilies of Lemuridae persist. The Lemurinae ("typical lemurs") and Lepilemurinae (Sportive lemur) are widespread, the Hapalemurinae (gentle lemurs) close to extinction.

The Lemuridae are squirrel- to cat-sized and weigh 1–11lb (0.5–5kg). Coat colors range from a muted gray-brown in the Sportive lemur and gentle lemurs to the striking black-and-white or black-and-red of the Ruffed lemur. Males and females are about the same size but in "true lemurs" (*Lemur* species) they differ in color, most strikingly in the Black lemur in which the male is jet black, and the female reddish- or golden-brown with lavish white tufts on the ears. All have long, often bushy, tails and have longer hindlimbs than forelimbs, a feature that is least pronounced in the Lemurinae (forelimbs measured at 69.7 percent of hindlimb length), somewhat more pronounced in gentle lemurs (67.1 percent), and most marked in the Sportive lemur (60.3 percent). These differences influence the way these generally arboreal animals move through the trees. "Typical lemurs" are largely quadrupedal; they move with agility amongst small branches and twigs on the periphery of tree crowns, and are capable of leaping several meters to cross from one crown to another. An exception is

▼ **Representatives** of the family Lemuridae ("typical lemurs") showing scent marking and sex differences in coat coloration of *Lemur* species. (1) Black lemur (*Lemur macaco*) male (ABOVE) and female. (2) Mongoose lemur (*L. mongoz*) male (ABOVE) and female. (3) Brown lemur (*L. fulvus*) marking its tail with the scent-glands on its wrist. (4) A subspecies, the White-fronted lemur (*L. fulvus fulvus*) male (the female has gray in place of white fur). (5) Gray gentle lemur (*Hapalemur griseus*) marking a branch with the scent glands on its wrist. (6) Sportive lemur (*Lepilemur mustelinus*). (7) Ruffed lemur (*Varecia variegata*) engaged in anogenital scent marking.

the Ring-tailed lemur, which habitually travels on the ground and, in the trees, prefers broad horizontal limbs to thin, less stable branches. It is the only primate in Madagascar to make extensive use of the ground, though the little known Crowned lemur may also be partially terrestrial. Gentle lemurs are not well studied, but have been observed moving quadrupedally and by leaping between vertical supports. The Sportive lemur travels almost exclusively by leaping from one vertical support to another; it rests by clinging to a tree trunk, sometimes in a fork but often with no horizontal support at all.

The muzzle is black, pointed, covered with sensitive whiskers and tipped with a naked, moist area of skin (rhinarium) that is linked to olfactory functions. Though all species, particularly the Sportive lemur, show some reduction in the area of the brain associated with olfaction, the sense of smell is still important: communication by smell is a conspicuous aspect of their behavior and all have scent glands which they use to mark certain branches or even one another. The Ring-tailed lemur's tail serves a double role in this respect. The black and white bands make it a striking visual signal, and during ritualized fights animals also smear their tails with secretions from the scent glands on their arms and wave them over their heads at the opponent. The family shares the typical primate trait of forward-facing eyes and binocular vision, though their binocular field is somewhat smaller (114–130°) than that of the monkeys (140–160°). Whether active by day or by

5

4

6

7

3

night, most Malagasy primates have a retina with a reflective layer (the *tapetum lucidum*), a specialization for night vision. Ruffed and "true" lemurs (except the Ring-tailed species) lack this characteristic, even though certain species of *Lemur* in some parts of their range are active by night as well as or instead of by day. Lemurids also communicate by voice. Ruffed lemurs and *Lemur* species use loud calls to draw attention to possible sources of danger and to maintain spacing between social groups; quiet calls help members of a group to stay in contact.

There are members of the Lemuridae in most of the remaining forests of Madagascar; Ring-tailed lemurs and perhaps Crowned lemurs occupy more open, scrubby areas in the south and north respectively. This is the only family of Malagasy primates with representatives in the wild outside Madagascar. The Mongoose lemur occurs on the Comoron islands of Moheli and Anjouan, and the Brown lemur on Mayotte. The date of their introduction (almost certainly by people) is unknown, but it goes back several hundred years. In Madagascar, the "true lemurs" and the Sportive lemur are found almost the length of the island in the east and the west, whereas gentle lemurs occur predominantly, and the Ruffed lemur exclusively, along the east coast. The largely deforested central plateau is devoid of primates today, but 1,000 years ago it contained a wide range of species, including representatives of all three lemurid subfamilies. In coastal areas the ranges of species and subspecies within each genus do not overlap, being broken up by stretches of uninhabitable terrain. The partial differentiation of recently isolated populations can make it difficult to decide whether they are subspecies or different species. The Sportive lemur presents a particular problem, and may be classified as one or as many as seven species. Differences in chromosome number have been used to separate populations of the Sportive lemur into species, but they are also found within populations in the same region. (This variation in chromosome number within a population, rare among primates, has also been observed in the Collared lemur.)

The family is vegetarian. They feed in a wide range of postures and can reach almost any part of a tree. Except for gentle lemurs, they rarely use their hands to manipulate food items but rather pull food-bearing branches to their mouth and feed from them direct. Particularly large items may be held while they are being eaten. Gentle lemurs have yet to be studied in detail, but they are reported to feed primarily on young bamboo shoots and leaves, and their front teeth appear specialized to exploit this food. The upper canine is short and broad and the premolar behind it is relatively large and not separated from the canine by a gap as it is in most primates. A gentle lemur will detach a bamboo shoot with its incisors, clamp it between the upper and lower canines and premolars, and pull the shoot sideways with its hands, stripping off the fibrous outer layer. It then pushes the tender interior back into the side of the mouth and chews it.

The Sportive lemur, studied in a spiny forest in the south, devoted 91 percent of its feeding time to leaves, and 6 percent to flowers and fruit together. Over half the food items came from just two plant species. This small, nocturnal animal may make the most of its low-energy diet by digesting its food by fermentation. It excretes the nutrients released by this process in its feces, eats the feces and assimilates the nutrients contained in them.

The diet of *Lemur* species varies from

The Bygone Wealth of Malagasy Lemurs

The evolution of the Malagasy lemurs is poorly understood because of the dearth of fossil evidence from the early and middle Tertiary in both Africa and Madagascar. One theory is that an early primate form, probably resembling in many respects today's Mouse lemur, colonized the recently formed island by clinging to rafts of vegetation as they swept out of river deltas draining eastern Africa some 50–60 million years ago.

Meeting no established mammalian competitors, these founding lemurs radiated to become an array of at least 40 species, evidence of which is preserved in sub-fossil deposits throughout western Madagascar. They spanned a wide range of physical and ecological types. Body size ranged from forms as small as a mouse to others as large as an orang-utan; all types of locomotion were represented: quadrupedal terrestrial, quadrupedal arboreal, slow grasping climber, vertical clinging and leaping, and brachiation (swinging by the arms); and, very likely, the larger-bodied types tended to be highly gregarious leaf-eaters, active chiefly in the daylight hours, like some monkeys in Africa today. The Indriidae alone, it is generally accepted, included (together with the present-day forms) the baboon-like *Archaeolemur*, the gelada-like *Hadropithecus*, the enormous and robust *Archaeoindris*, and the large brachiating *Palaeopropithecus*. Competing with *Archaeoindris* in body size was *Megaladapis*

(ABOVE), koala-like in proportions but with a skull over a foot (25cm) long in one species, proving that the evolutionary diversification involved more than a single family.

There is good evidence that the majority of these large lemurs were still widespread up to about 5,000 years ago and were subsequently devastated by hunting, with the arrival of the first permanent human settlements in Madagascar about 2,000 years ago. Fire and competition with domestic animals may also have accelerated their disappearance. Whichever the case, we have just missed a unique opportunity to observe the extraordinary capacity of evolutionary processes to construct variations on a theme, with the primates isolated on Madagascar evolving in parallel with the monkeys and apes of the Old World. JIP

▲ **The unmistakable tail** of the Ring-tailed lemur is used by this day-active species as a visual signal. In aggressive encounters, the lemur will wave high its scent-smeared tail in the direction of its rival.

◀ **The Mongoose lemur** in Madagascar lives in small groups of male, female (shown here) and young offspring. In the Comoro Islands this nectar-eating lemur lives in larger groups.

▶ **In mid-leap** OVERLEAF, a Ring-tailed lemur prepares to land on an upright branch.

species to species and, within species, according to region and season. For example, the Mayotte lemur eats primarily fruit (67 percent of feeding time) and leaves (27 percent) from a wide range of species, but the proportion of these plant parts in its diet varies seasonally from 48 to 79 percent for fruit and 20–36 percent for leaves. The Red-fronted lemur, in contrast, spends over half its feeding time on the leaves of a few species. The diet of the Mongoose lemur contains a unique component: during the dry season, animals spend 81 percent of feeding time licking nectar from flowers or eating the nectaries themselves.

Social organization and ranging behavior vary widely within the family, and in *Lemur* there is variation within as well as between species. In Madagascar, Mongoose lemur social groups consist of an adult male and female with their immature offspring. They live in small (about 3-acre/1.5ha) overlapping home ranges, and the bond between the two adults may last several years. Among Mongoose lemurs in the Comoro Islands, however, there is no evidence of pair-bonding; animals live in larger groups of highly variable composition. In Madagascar, social groups of Red-fronted lemurs contain about nine animals (range 4–17) and occupy an extremely small (about 2-acre/0.75–1ha) home range that overlaps extensively with the home ranges of neighboring groups. Groups are stable in composition but, other than grooming, interactions among members are rare and there is no clear social structure. In the Comoro Islands, Mayotte lemurs also spend their time in groups of about nine animals, but the composition of these groups varies from day to day, and they may be subunits of a larger social network. Social group size in the Ring-tailed lemur varies from five to 30 (average 15). The home range is larger than in other lemurs (15–57 acres/6–23ha), and it may be defended and used exclusively, or partly shared with neighboring groups. Aggressive interactions within the social group are common and there are clear male and female dominance hierarchies. Females are dominant to males. The male hierarchy disintegrates during the mating season and males compete equally for access to females. Females spend their lives in the group into which they were born, whereas males transfer at least once and possibly several times during their lives from one group to another. The Ruffed lemur has yet to be well studied but it appears to live in pair-bonded units. Its pattern of reproduction is unique in the family. The female usually gives birth to twins, and during the postnatal period she "parks" them on a branch or in a nest while she forages.

In contrast to the sociable "typical lemurs," the Sportive lemur spends most of its time alone, though heterosexual or female pairs often share most or all of a tiny home range (1 acre/0.3ha for males and 0.5 acre/0.18ha for females). Meetings occur 1–3 times a night, lasting from five minutes to an hour, during which the pair move, feed, rest and occasionally groom together. The social organization of gentle lemurs has not been studied; however, they have been observed in groups of two to six animals.

All the Lemuridae are threatened by habitat destruction, but differences in ecology may affect the length of time members of the three subfamilies will be able to survive on the island. The "typical lemurs" eat a wide variety of foods; their ranging behavior is flexible and some species are able to be active by day or by night or even both; their social organization also seems to be adaptable. They live at high densities in favorable habitats and occupy many types of vegetation over much of the island. The diet of the Sportive lemur in a given forest is likely to be much narrower than that of the "typical lemurs," but these small, solitary, nocturnal and inconspicuous animals are specialized to exploit one of the most abundant resources in the forest, namely leaves. They are thus able, like "true lemurs" though for different reasons, to survive at high densities in many forests and are widespread on the island. Gentle lemurs, by contrast, are specialized eaters of bamboo shoots, a plant that grows only in a limited range of environmental conditions. Gentle lemurs are thus much less common, their distribution is patchy, and they occur only in the island's moister forests. AFR

Dwarf and Mouse Lemurs

The nocturnal dwarf and mouse lemurs (family Cheirogaleidae) are the smallest Madagascan primates, the highest body weight being about 1.1lb (500g). Like all lemurs (except the Ring-tailed lemur) they spend virtually all their time in the trees. They only descend to the ground to cross large gaps between trees or, in the case of the mouse lemurs, to catch terrestrial beetles and other arthropods in a rapid dash before returning to the comparative safety of the trees.

Dwarf and mouse lemurs can be loosely described as "omnivorous," since all species consume both animal prey (mainly insects) and plant food, but each species tends to have a dietary speciality. The family is accordingly widespread in Madagascar, the most common species being the two lesser mouse lemurs, particularly adaptable forest-fringe inhabitants which occupy even the tiniest patches of natural forest. The remaining species have more restricted distribution.

In common with most other lemurs and with the Afro–Asian bush babies and lorises (p36), mouse and dwarf lemurs possess a reflecting *tapetum* behind the retina of each eye, a flat layer containing plate-like riboflavin crystals which assist vision in dim light by shifting the wavelength of incident light to the most sensitive range (yellow) for the light receptors, and by reflecting it back through the retina. For field observers, the reflections from the eyes facilitate location of these animals at night with the aid of a headlamp.

Apart from the little-known Hairy-eared dwarf lemur, the smallest members of the family are the two species of lesser mouse lemurs. Their body weight, which averages about 2oz (60g), fluctuates markedly during the year, since fat reserves are stored in the tail during the wetter months and then used up to offset reduced availability of food during the drier months. Both Gray and Brown lesser mouse lemurs use nests, sometimes constructing them as spherical balls of leaves wedged between branches and sometimes making use of natural tree hollows. Active animals can be seen out in the trees in all months of the year, though some individuals may become somewhat torpid during the drier months, particularly in response to low temperatures. Lesser mouse lemurs have the most generalized diet of all the lemurs. Their staple food consists of small fruits, but they are known also to eat various arthropods and small vertebrates as well as feeding occasionally on natural exudates of gum.

▲ **Storing fat in its tail,** the slow-moving, Fat-tailed dwarf lemur clambers from branch to branch and remains dormant for half the year.

► **The Brown lesser mouse lemur** OPPOSITE TOP, one of the smallest primates, leaps from tree to tree. Large, forward-facing eyes pick out a safe landing place in the forest night.

Coquerel's mouse lemur is considerably larger than the two other *Microcebus* species and has a far more restricted distribution. Its diet similarly includes a fair proportion of small fruits and arthropods, but it has at least one dietary peculiarity: colonies of flattid bugs are found on small trees and bushes in western Madagascar at the end of the dry season, when Coquerel's mouse lemurs spend much time licking the secretions produced by these bugs. During the day Coquerel's mouse lemurs sleep in globular leaf nests.

In contrast to the actively leaping mouse lemurs, the two species of dwarf lemurs are slow and deliberate in their movements, usually moving directly from branch to branch. Because of their resulting need for adjoining trees, dwarf lemurs are not as common in Madagascar as the lesser mouse lemurs. Whereas the eastern and western species of lesser mouse lemur are very similar in body size, the Greater dwarf lemur of the east coast rain forests weighs more than twice as much as the Fat-tailed dwarf lemur of drier western forests. Both species store fat in their tails, though this is more pronounced in the smaller Fat-tailed dwarf lemur, and both are truly dormant for at least six months during the drier months (April–October). Dwarf lemurs are more exclusively fruit-eaters than lesser mouse lemurs, though they do eat some arthropods (mainly beetles) and gum in addition.

Virtually nothing is known about the small Hairy-eared dwarf lemur, which is only documented by a few preserved specimens. However, its "needle claws" and certain features of the dentition indicate that this species is specialized for gum-feeding.

The final member of the family, the Fork-crowned dwarf lemur, is an actively leaping, hole-nesting species of relatively large body size. Approximately 90 percent of its diet consists of gum exudates collected from the surfaces of trunks and branches, the remaining 10 percent being mainly arthropods. Numerous anatomical adaptations reflect this dietary specialization. The tooth-comb formed by the forward-projecting (procumbent) lower incisors and canines is noticeably elongated for scraping off the gums. The nails have sharp tips ("needle claws") for clinging to broad trunk surfaces during gum-feeding, while the cecum (a blind off-shoot of the gastrointestinal tract) is enlarged and houses symbiotic bacteria essential for the digestion of the gums. Perhaps because of its specialized diet, the Fork-crowned lemur is found only in certain areas of Madagascar's coastal forests. It is conspicuously absent from most of the east coast rain forest.

All mouse and dwarf lemurs are "solitary" in that they generally feed alone at night, but detailed observation has revealed quite complex social networks, commonly based upon range overlap between several females and a single fully adult male (extended harem system). There is some evidence that Coquerel's mouse lemur may tend towards pair-living, and the Fork-crowned lemur definitely seems to be monogamous, the male and female traveling in close proximity and maintaining continuous vocal contact. Vocalizations generally play a major part in spacing, though this is the only species with clearly defended, exclusive territories.

Dwarf lemurs and lesser mouse lemurs have two or three infants rather than the singletons typical of most primates. All members of the family keep their infants in the nest until they are capable of independent feeding. Like virtually all other mammals in Madagascar, mouse and dwarf lemurs breed seasonally. Most species have one litter a year, but lesser mouse lemurs may produce two, particularly if the first is lost. The breeding season generally seems to be timed so that the young can build up adequate food reserves to survive the drier months that follow. RDM

Indri and Sifakas—the leaping lemurs

Only three genera of this once extensive family (Indriidae) of large-bodied lemurs survive today. The recent extinction of about 50 percent of Madagascar's diverse lemur fauna reduced these actively leaping lemurs to just four representatives, three of which typify the family as a whole in being large, leaf-eating lemurs active in daytime and living in small groups. One of these, the indri, largest of all prosimians, is the "babakoto" that figures prominently in Malagasy accounts of the origin of man.

The indris and sifakas are monkey-sized lemurs which move beneath the canopy of trees by leaping from one vertical trunk or bough to another with the body usually in an upright position. They are almost wholly arboreal and visit the ground only for eating

The Aye-aye, Strangest of Primates

The aye-aye has a unique appearance and habits. It has a thick, dark brown coat, white-flecked on the body, with exceptionally long guard hairs, a big bushy tail, naked, bat-like ears, huge incisors (at least 0.4–2.4in/1–6cm long) separated by a considerable gap (at least 0.5in/1.25cm) from the few other teeth (dental formula I1/1, C0/0, P1/0, M3/3 = 18), a spindly middle finger on each hand, and claws on all digits (which are very elongated) except the big toes, which have nails. At 6.6lb (3kg) and with head-body and tail length both about 16in (40cm), it is also the largest nocturnal primate. Much remains to be learned about it.

The aye-aye lives only in the northern sector of the east coast rain forests of Madagascar. A solitary creature, by day it sleeps in a nest which it may occupy for several days before building a new one. By night it moves quadrupedally through the trees or on the ground. In the trees, its claws help it cling to trunks and to very thin branches; it can even hang upside down on such branches, using its claws as hooks. On the ground, the fingers are raised so that they do not touch the ground, giving the animal a strange and clumsy gait.

The aye-aye eats fruit and insect larvae. It uses its powerful, forward-curving bevel-edged incisors to tear through the hard outer shell of fruits such as the coconut or to scrape away at fibrous fruits like the mango. Uniquely among primates, the aye-aye's incisors are continuously growing to allow for heavy wear at their tips. Once a hard-shelled fruit has been pried open, its inner pulp and juices are extracted with the middle finger. The aye-aye's large ears probably help in the detection of larvae hidden under the bark of dead branches; once found, the larvae are exposed by ripping off part of the overlaying bark with the incisors. The aye-aye probes the hole thus made with its middle finger and transfers the larvae to its mouth. Madagascar has no woodpeckers, and aye-ayes may have evolved to fill the predatory role played elsewhere by these birds.

The aye-aye is the only surviving member of the family Daubentoniidae. It is now almost extinct. A similar species with dimensions one-third larger, *Daubentonia robusta*, became extinct in the last 3,000 years. Habitat destruction is an important cause of their disappearance, but people have also had a more direct hand in it. Prompted perhaps by their bizarre appearance, many Malagasy consider them creatures of ill-omen, to be killed on sight. AFR

bark or earth (see below) or to cross small treeless areas. The indri may leap up to 33ft (10m), although most leaps are of 10–16ft (3–5m). Movement on the ground is also unique: bipedal hops with arms held at shoulder level or higher, the body inclined somewhat away from the direction of travel.

There are few physical distinctions between the three genera; they include the relatively small size of the Woolly lemur (the only nocturnal species), the near-absence of a tail in the indri, and body coloration differences. Common indriid features include a high leg-to-arm-length ratio (an adaptation to their mode of locomotion), the loss of one premolar tooth from both upper and lower jaws, four rather than six teeth in the tooth-comb, and a single pair of axillary teats. They have short, broad snouts and large hands and feet. There have been field studies of Verreaux's sifaka and of the indri, but little is known of the Diadem sifaka and Woolly lemur.

The diet of indriids is composed entirely of fruit and leaves. A small amount of bark and dead wood is consumed by Verreaux's sifaka (probably to obtain water) and the indri also ingests earth (possibly as an aid to digestion). Indriids generally feed by pulling branches manually to the mouth and biting off the food items. Their grasping hands and feet enable them, despite their size, to hang from thin branches to reach food items. Over the year, Verreaux's sifaka spends about 43 percent of its feeding time eating young and mature foliage, 38 percent on fruit and the remainder on flowers, bark and dead wood. The proportion of time spent feeding (24–37 percent) varies seasonally. Feeding accounts for 35–40 percent of the indri's daily period of activity on average and 60–70 percent of this is spent eating leaf shoots and young leaves, 25–30 percent on fruit and seeds and the remainder on flowers and mature foliage. All indriids, especially the indri, possess a well-developed cecum where plant cellulose is presumably fermented with the aid of gut microflora.

Groups of Verreaux's sifaka are composed of 3–12 individuals, often with more than one breeding adult of each sex. Some males transfer between groups before the mating season, apparently according to their social status within the group. Females can reproduce about three years of age and have a very short (up to 42 hours) period on heat in late summer (February–March) which is highly synchronized within and between groups and results in a sharp peak in the birth rate in June and July after a 160-day gestation. The single newborn infant is almost hairless and black-skinned, clutching the lower part of the mother's abdomen to begin with but gradually transferring to

▲ ► **The long, muscular legs,** large grasping hands and feet, and long balancing tail of an active leaper. This white Verreaux's sifaka ABOVE on a cactus-like *Didiera* is one of the largest lemurs. Like most larger species it is active during the daytime.

◄ **Limbs outstretched,** this indri's midair posture is that of a "vertical clinger and leaper," the largest in fact, and the largest of all prosimians.

her back after 3–4 weeks. Weaning is completed by five months and the young sifaka moves independently of the mother from about seven months.

Indri generally live in small family groups, with only one breeding adult of each sex. Mating occurs in mid-summer and single infants are born in May after a gestation of probably 120–150 days. Compared with the sifaka, development is slow: the infant is carried on its mother's back from 15 to about 28 weeks and moves about independently at 42 weeks, some three months before weaning. Births probably occur at three-year intervals, and reproductive maturity is attained at the age of 8–9 years.

Verreaux's sifaka occupies ranges of 2.5–22 acres (1–9ha) in groups of 1–4 adult females, 1–4 adult males and young of both sexes, with a mean group size of 5.9. In larger ranges, a quarter or half the range is an actively defended territory—in smaller ranges the whole area may be territorial. Population densities of over 250 per sq mi (100/sq km) occur in this species, much more than in the Diadem sifaka, which lives in largely exclusive ranges of at least 49 acres (20ha) in groups of only 2–5 individuals. In both sifaka species females engage in ano-genital scent-marking and males rub glands in the throat region against trees and branches—behavior which increases in frequency in the mating season. Indri groups defend territories of 44–74 acres (18–30ha); population density is estimated at 23–41 per sq mi (9–16/sq km) where the species is abundant. Indri males "cheek mark," and ano-genital marking is performed by both sexes.

Indri may also defend their territory by the groups' loud, modulated series of howls, which occur regularly in infectious fashion throughout the population. These "songs" are emitted 1–7 times a day on 70 percent of days during the year. They serve to recall separated group members and are repeated during the occasional encounters between groups at territory borders. Sifakas also occasionally emit very loud "football-rattle" calls and both the indri and sifakas call distinctively to warn of ground predators ("hoot"—indri; "sifaka"—*P. verreauxi*; "kiss-sneeze"—*P. diadema*) and aerial predators ("roar"—indri; and "football-rattle"—sifakas). Soft calls within the family group consist of hums, grunts and growls. Each species appears to have its own distinct repertoire of 5–6 calls.

Woolly lemurs live in small groups of 2–5 but are frequently encountered alone. Four or five cries, some loud, have been described. Neck glands are developed in both sexes.

All indriids are severely threatened by the destruction of their habitat for fuel, timber or local agricultural development. The eastern species are in most danger because the rain forest of eastern Madagascar is becoming fragmented into a series of thin islands, which is certain to result in extinctions. Both species of sifakas are eaten locally and are commonly trapped and shot. Only Verreaux's sifaka has been satisfactorily maintained in captivity. The few remaining species of the family can only be protected by efficient management, and perhaps by extensions to the 15 protected areas, which currently cover just 1 percent of Madagascar. JIP

THE 22 SPECIES OF LEMURS

Abbreviations: HBL = head-and-body length; TL = tail length; wt = weight.
[E] Endangered. The status of all lemurs is under review by the IUCN.

Lemurs (family Lemuridae)

Subfamily Lemurinae (typical lemurs)

Genus *Lemur* ("true" lemurs)

Medium-sized primates (4–7lb). Mostly arboreal, with a diet of fruit, leaves, and flowers. Somewhat prominent muzzle, tipped with a moist rhinarium. Dental formula I2/2, C1/1, P3/3, M3/3 = 36. Lower canines and incisors form a forward-projecting "comb." Arms shorter than legs, and tail longer than body. Usually move on all four limbs. Often striking male/female differences in coat color. Sexual maturity at about two years of age. Mating seasonal. Usually a single offspring after a gestation of 120–136 days. Several species have received little study.

Ring-tailed lemur

Lemur catta

Madagascar. Deciduous forests, gallery forests, *Euphorbia/Didierea* arid bush/forest. Diurnal. Gregarious, in groups of 5–30 with both male and female dominance hierarchies, but females generally dominant to males. Feeding largely arboreal, but animals often travel on ground. Births in August–November (N Hemisphere, March–June). Coat: back gray, limbs and belly lighter, extremities white; top of head, rings about eyes and muzzle black; tail banded black and white. HBL 15–18in; TL 22–25in; wt 5–8lb.

Lemur fulvus

Seven subspecies from E and W coasts of Madagascar and from Comoro Islands. In all types of forest, except in dry S and SW. Highly arboreal with a preference for closed-canopy forests. Little evidence of dominance hierarchies. May live at very high densities (up to 5 per acre). Group size variable (4–17). Some populations diurnal, others sporadically active through day and night. Some of 7 subspecies becoming very rare. Sometimes included in *L. macaco*.

Brown lemur

L. f. fulvus

Range discontinuous, along NW coast NE of Galoka mountains, and central E coast of Madagascar. Upper parts and tail grayish-brown, cheeks and beard white, muzzle and forehead black, underparts creamy-tan; color paler in females. HBL 17–20in; TL 20–22in; wt 5–9lb. Groups 3–12 individuals, predominantly diurnal (?).

White-fronted lemur

L. f. albifrons

E coast rain forests, N of *L. f. fulvus*. Coat color variable, different between sexes. Males usually brown on upperparts and tail, paler underneath with striking black face and muzzle surrounded by snowy white forehead, crown, beard and throat. Females have gray-brown backs and tails, with lighter underparts; muzzles black, but otherwise heads usually grayish. HBL 15–16in; TL 20–22in; wt 4–6lb.

Red-fronted lemur

L. f. rufus

Forests along much of W coast (S of *L. f. fulvus*) and on central E coast (S of *L. f. fulvus* and N of *L. f. albocollaris*). Coat colors variable, different between sexes. Male upperparts and tail gray, underparts lighter; muzzle black, head hooded in reddish-orange, cheeks and throat grayish-brown. Female upperparts reddish-brown, underparts lighter or grayer; muzzle and center of forehead black, crown gray with pale eyebrows and cheeks. HBL 15–18in; TL 20–24in; wt 5–8lb. Groups of 4–17 animals with small overlapping home ranges. High proportion of mature leaves in diet, also fruit, few flowers. Feeds throughout day with frequent rests. Aggression very rare, dominance hierarchies nonexistent (?). Most abundant subspecies.

Collared lemur

L. f. collaris

Extreme S section of E coast rain forest. Upperparts olive-brown to gray-brown, with darker strip down spine. Underparts paler, tail darker. Face and top of head black, cheeks, orange and bushy in males. Females similar but face grayer, and cheeks have shorter hair. HBL ? TL ? wt 5–6lb.

Mayotte lemur

L. f. mayottensis

Limited to island of Mayotte, in Comoro Islands NW of Madagascar. Coat very similar to *L. f. fulvus*, but individual variation somewhat greater. HBL about 16in; TL about 20in; wt? Closely to *L. f. fulvus*. Likely a human introduction to Mayotte. Highly arboreal, active chiefly in early morning and evening, but sporadically at all times. Group size variable (2–29). Aggression rare, dominance hierarchies not apparent.

Sanford's lemur

L. f. sanfordi

Limited to vicinity of Mt d'Ambre in far N of Madagascar. Male upperparts brownish-gray, underparts paler, limbs sometimes reddish; black muzzle, with forehead and upper cheeks whitish; crown and lower cheeks bushy and reddish-brown, ears white and tufted. Females have similar body colors but dark-gray head, no ear tufts, less bushy cheeks. HBL 14–17in; TL 18–20in; wt ?

White-collared lemur

L. f. albocollaris

S east coast rain forest, between *L. f. collaris and L. f. rufus*. Males similar to *L. f. collaris* males, but beard is white. Females similar to *L. f. collaris* females. HBL 16–17in; TL 18–23in; wt ?

Mongoose lemur

Lemur mongoz

NW Madagascar, and Moheli and Anjouan in Comoro Islands. Moist forest, deciduous forest and second growth. Likes nectar, also eats flowers, fruit and leaves. Most populations nocturnal, arboreal, and live in family groups (male, female and immature offspring). Aggressive behavior uncommon. Males gray, with pale faces, red cheeks and beards. Females have browner backs, dark faces, white cheeks and beards. HBL 12–14in; TL 18–20in; wt 4–5lb.

Black lemur

Lemur macaco

West of N Madagascar. Humid forests. Primarily arboreal, daytime and dusk activity, foraging groups of 5–15 animals may coalesce in evenings. Highly dichromatic: males uniformly black; females light-chestnut brown, with darker faces and heavy white ear tufts. HBL 15–18in; TL 20–25in; wt 4–6lb.

Crowned lemur

Lemur coronatus

Extreme north. Dry forests and, recently, high-altitude moist forest. Apparently lives in large, multi-male groups. Active in day and at dusk. Travels frequently on ground. Sexually dichromatic. Males have medium-gray backs, lighter limbs and underparts; faces whitish, with V-shaped orange marking above forehead; crown of head black. Female upperparts and head cap lighter in color. HBL 12–14in; TL 16–20in; wt about 4lb.

Red-bellied lemur

Lemur rubriventer

Medium to high altitudes in E coast rain forest. Poorly known but probably diurnal, limited to highest strata of forest, always in small groups of up to 5. Upperparts chestnut-brown, tail black, face dark. Males have reddish-brown underparts, females whitish. HBL 14–16in; TL 18–21in; wt ?

Genus *Varecia*

Ruffed lemur

Varecia variegata

Sparsely distributed in E coast rain forest. Largest of Lemuridae. Poorly known: probably fruit-eating, in upper strata of forest. Live in pair-bonded families; high rate of twin births. The only lemurid that leaves young in nest. At least 2 subspecies recognized, primarily on basis of coat color. Both have long, dense fur, especially around neck. Prominent muzzle, face covered by short hair.

Black-and-white ruffed lemur

V. v. variegata

Central and S portion of E coast rain forest. Geographically highly variable coat colors; usually white ruff with black face, extremities and tail; shoulders, back and rump have black and white patches and bands of various extent. HBL 21–22in; TL 23–26in; wt 7–10lb.

Red ruffed lemur

V. v. rubra

N part of species' range. Body deep red, except underparts: extremities, forehead, crown and tail black, white patch at the back of the neck. HBL 20–22in; TL 22–24in; wt about 9lb. Exceptionally rare.

Subfamily Lepilemurinae (Sportive lemur)

Sportive lemur

Lepilemur mustelinus

Xerophytic, gallery, deciduous and humid forests of Madagascar. Medium-sized (HBL 9–12in, TL 9–11in, wt about 1–2lb). Arboreal, nocturnal, usually sleep during day in tree hollows. Diet largely leaves, also fruit, flowers, bark. Move by leaps of up to 16ft from trunk to trunk with body in vertical position. Legs much longer than arms, tail always shorter than body. Short face, moist rhinarium, prominent ears, dense, woolly fur. Dental formula I0/2, C1/1, P3/3, M3/3 = 32; lower incisors and canines form a forward-projecting tooth-comb. Single births September–November after gestation of 135 days; sexual maturity attained at about 18 months. Six subspecies.

Weasel lemur
L. m. mustelinus
E coast rain forest. Upperparts brown, underparts paler, face dark, lighter cheeks and beard. Perhaps largest subspecies.

Red-tailed sportive lemur
L. m. ruficaudatus
Didierea/Euphorbia bush, gallery forests in SW. Back light brown, underparts paler, reddish tail, pale face and throat. Large ears.

Gray-backed sportive lemur
L. m. dorsalis
NW, moist forests. Small-bodied. Medium to dark brown above and below, face dark, ears small.

Milne-Edward's sportive lemur
L. m. edwardsi
W deciduous forests. Similar to *L. m. ruficaudatus*, but coat darker, especially on upper part of back. Gray-brown face, underparts gray, ears large.

Northern sportive lemur
L. m. septentrionalis
Extreme N, deciduous forests. Upperparts and crown gray, rump and hindlimbs paler, tail pale brown. Underparts and face gray.

White-footed sportive lemur
L. m. leucopus
S gallery forests and *Didierea/Euphorbia* bush. Small-bodied. Upperparts medium gray, underparts gray-white, tail light brown. Ears large.

Subfamily Hapalemurinae (gentle lemurs)

Genus *Hapalemur*
Madagascar. Similar in size to *Lemur*. Active morning and late afternoon. Specialized diet of bamboo shoots and reeds. Moves quadrupedally, as well as by leaps, often near the ground. Rounded head, with small, furry ears, short muzzle and woolly coat. Dental formula I2/2, C1/1, P3/3, M3/3 = 36, with upper canine and P2 specialized for unsheathing bamboo shoots. Forward projecting tooth-comb. Single births after 140-day gestation. Two living species.

Hapalemur griseus
E coast humid forest and two isolated populations along W coast. Essentially limited to bamboo forests and reed beds. Three subspecies.

Gray gentle lemur
H. g. griseus
Throughout eastern humid forest. Small family (?) groups of 2–6 animals. Coat largely gray to gray-brown with lighter underparts. HBL 7–12in; TL 12–16in; wt 1.5–2lb.

Alaotran gentle lemur
H. g. alaotrensis
Only in reed beds and marshes surrounding Lake Alaotra, E Madagascar. Coat similar to *H. g. g.* HBL about 16in; TL about 16in; wt probably about 2lb.

Western gentle lemur
H. g. occidentalis
Two small isolated populations in bamboo forests along W coast. Somewhat smaller and lighter colored than *H. g. g.* HBL about 11in; TL about 15in; wt probably about 2lb.

Broad-nosed gentle lemur
Hapalemur simus
Extremely limited range in central E coast humid forest. Coat gray to gray-brown with lighter underparts. Larger and more heavily built than *H. griseus*. Total length (HBL + TL) about 35in. In immediate danger of extinction.

AFR

Dwarf and Mouse Lemurs (family Cheirogaleidae)

Fat-tailed dwarf lemur
Cheirogaleus medius
NW, W and S Madagascar in well-established secondary forests and primary forests; nocturnal. HBL 7in; TL 8in; wt 7oz. Fur short and dense; pale gray above, white to cream below. Gestation: 61 days. Longevity: 18 years in captivity.

Greater dwarf lemur
Cheirogaleus major
E Madagascar in well-established secondary forests and primary forests; nocturnal. HBL 9in; TL 11in; wt 16oz. Fur fairly short and dense; gray-brown above, white to cream below; black ring around each eye. Gestation: 70 days. Longevity: 15 years in captivity.

Gray lesser mouse lemur
Microcebus murinus
NW, W and S Madagascar, in forest fringes and secondary vegetation; nocturnal. HBL 5in; TL 5in; wt 2oz. Fur on back gray to gray-brown; white to cream below. Large membranous ears. Gestation: 60 days. Longevity: 14 years in captivity.

Brown lesser mouse lemur
Microcebus rufus
E Madagascar, in forest fringes and secondary vegetation; nocturnal. HBL 5in; TL 6in; wt 2oz. Fur on back brown; white to cream below. Medium-sized membranous ears. Gestation: 60 days. Longevity: 12 years in captivity.

Coquerel's mouse lemur
Microcebus coquereli
Disjunct distribution in well-established coastal forests of W and NW Madagascar; nocturnal. HBL 8in; TL 13in; wt 10oz. Fur on back gray-brown to pale brown, yellowish below. Tip of tail darker than rest of fur. Large membranous ears. Gestation: 86 days. Longevity: unknown.

Hairy-eared dwarf lemur
Allocebus trichotis
NE Madagascar; very restricted distribution in primary rain forests; nocturnal. HBL 5in; TL 6in; wt unknown, probably about 2oz. Fur on back pale brown; white to cream below; ears short but with pronounced tufts of long hair. Gestation and longevity unknown.

Fork-crowned dwarf lemur
Phaner furcifer
Disjunct distribution in W, NW and NE Madagascar, in well-established forests in coastal regions; nocturnal. HBL 9in; TL 14in; wt 10oz. Fur gray-brown on back, white to cream below; conspicuous dark line running over back of head and forking to join up with dark rings surrounding the eyes. Gestation and longevity unknown.

RDM

Indri and Sifakas (family Indriidae)

Woolly lemur
Avahi laniger
Woolly lemur, woolly indris or avahi.
Two subspecies, in E rain forests, also in secondary growth (*A. l. laniger*) and central NW coastal area (*A. l. occidentalis*). HBL of *A. l. l.* 10–12in; TL 11–14in; wt 2lb. Coat thick, dark, usually with white inside thigh patches.

Indri
Indri indri
Indri or indris, endrina or babakoto.
Rain forest along N central part of E coast. HBL 22–27in; rudimentary tail 2in; wt 15–22lb or more. Coat variegated black and white, thick, silky; amount of white variable.

Verreaux's sifaka
Propithecus verreauxi
Four subspecies, in deciduous forests of S, W and N. HBL 15–19in; TL 20–24in; wt 8–9lb. Coat varies greatly from all-white to partly or largely brown, black or maroon.

Diadem sifaka
Propithecus diadema
Five subspecies, in evergreen forests in E. HBL of *P. d. candidus* 20–22in; TL 18–20in; wt 13–18lb. Coat varies from all-white to all-black, with extensive gold, gray or brown patches in some subspecies.

JIP

Aye-aye (family Daubentoniidae)

Aye-aye [E]
Daubentonia madagascariensis
E coast rain forests. HBL about 16in; TL about 16in; wt 7lb. Coat thick, dark brown, white-flecked on body, with long guard hairs. Tail bushy. Incisors very large. Fingers and toes very large; thin middle finger. Nocturnal and solitary; eats fruit and insect larvae. (See p31.)

BUSH BABIES, LORISES AND POTTOS

Family: Lorisidae.
Ten species in 5 genera. (See table of species p40.)
Distribution: warm regions of Africa and Asia.

Habitat: rain forest to dry forest and savanna with trees.

Size: mostly weighing 7–10.6oz (200–300g), but ranging from head-body length 4.7in (12cm), tail length 10.6in (27cm) and weight 2.1oz (60g) in the Dwarf bush baby to head-body length 12.6in (32cm), tail length 17.3in (44cm) and weight 2.6lb (1.2kg) in the Thick-tailed bush baby (potto and Slow loris similar in body length and weight).

Gestation: 110–193 days.

Longevity: In species for which information is available, ages attained in captivity range from 12 to 15 years or more.

Bush babies (subfamily Galaginae)
Six species in Africa, four more or less limited to west equatorial rain forests: the **Dwarf bush baby** (*Galago demidovii*) or Demidoff's galago, **Allen's bush baby** (*G. alleni*), and two **Needle-clawed bush baby** species (*G. elegantulus* and *G. inustus*). Two species in more arid regions from Senegal to E. Africa and down to southern Africa: the **Lesser bush baby** (*G. senegalensis*) and the **Thick-tailed bush baby** (*G. crassicaudatus*).

Pottos and lorises (subfamily Lorisinae)
Four species, two in Africa: the **angwantibo** (*Arctocebus calabarensis*) or Golden potto, with very limited distribution in W equatorial regions, and the **potto** (*Perodicticus potto*), whose distribution extends considerably further both E and W. Two species in Asia: the **Slender loris** (*Loris tardigradus*) in tropical forests of India and Sri Lanka, and the **Slow loris** (*Nycticebus coucang*), from Vietnam to Borneo.

A crude definition of the family that includes the bush babies, pottos and lorises would be that they are lemurs that do not inhabit Madagascar. Unlike their Madagascan counterparts, these continental species share their habitat with monkeys. However, because their activity is exclusively nocturnal the Lorisidae do not compete ecologically with the monkeys.

Like the lemurs, lorisids have retained a well-developed sense of smell and can be distinguished from the higher primates (and the tarsiers—see p42) by the presence of a moist snout (rhinarium). In addition, unlike the higher primates, their face is covered with hair. The dental formula (I2/2, C1/1, P3/3, M3/3 = 36) is almost the same as in the earliest lemur-like creatures of 50 million years ago, although modern members of the family, like most of the Madagascan lemurs, are characterized by the presence of a tooth-comb formed by the four lower incisors and two canines, which are pointed and project forward. Species that feed on gum and other plant exudates use the comb to scoop out drops held in fissures of the tree bark. During grooming the comb is used to remove any rough material, encrusted mud, tangled hair etc. On the underside of the tongue lies a second, fleshy comb (the

► **Bat-like ears** of the Lesser or Senegal bush baby enable it in the dark to track movements of its insect prey. Some prey (eg arthropods) is taken on the ground, but insects may be snatched as they fly past.

▲ **The nocturnal Slender loris** has large eyes and binocular vision. Not unlike bush babies, lorises and pottos are slow movers and can "freeze" chameleon-like for hours. They locate their slow-moving prey particularly by smell. The Slender loris of India and Sri Lanka is one of just two Asian members of this chiefly African family.

▼ **Thick-tailed bush babies** hang from a branch as they feed on gum trickling down the trunk of an acacia.

sublingua). This has sharpened and hardened points used to clean the debris from between the teeth of the dental comb.

The fingers and toes all bear nails, except the second toe which is modified as a toilet claw and used to groom the head and neck fur and to clean the ears. Only the first toe of the foot is truly opposable to the other digits.

As in all primates the external sexual organs are clearly visible. However, the clitoris is so developed as to lead to possible confusion between the sexes, especially as the vaginal opening is nearly always obscured by the growth of "scar tissue" between the female's fertile periods. The surest method of sexing individuals is to determine whether a scrotum is present or not.

The best known members of the family are the bush babies of Africa—so called perhaps on account of their plaintive cries and also because of their cute appearance—and the lorises of India and Sri Lanka, which take their name from the acrobatic and "comical" postures adopted as they move about ("loris" means clown in Dutch, language of the seafarers who first brought individuals to Europe).

The different ways in which members of the group move about are one of its most remarkable characteristics. Two subfamilies are recognized, based on modes of locomotion developed along two diametrically opposed evolutionary paths. The bush babies (subfamily Galaginae) are agile leapers, and comprise six species, all in Africa. The bush babies' leaping and fast-running progress is equally effective in the thick foliage of the rain forest and in the trees of the savanna. With a series of leaps between branches or tree trunks a bush baby can cover some 33ft (10m) in less than five seconds, and can effectively evade an intending predator.

Several morphological adaptations are linked to this type of locomotion. The hind limbs are highly developed, the eyes are large for good night vision, and the long tail is important for keeping balance during a leap.

The pottos and lorises (subfamily Lorisinae) are slow climbers. There are two African and two Asiatic species. These four species have a much reduced tail and limbs of less unequal length than in the bush babies. They look like slow and cautiously moving bear cubs. They move along a branch or a liana in a somewhat chameleon-like movement that is smooth and perfectly coordinated, so that they remain unnoticed as they pass through the thick vegetation. This system of concealed or "cryptic" locomotion has been developed to such a degree that they have lost the skills of leaping. One captive potto remained amidst some branches fixed to the ceiling of a room for two years without once falling to the floor 6.6ft (2m) beneath the lowest branches. Only in the case of intense fear, for example when confronted with a large snake, will these animals let themselves drop to earth, an effective means in thick forest of escaping. When a potto or loris is on the move, it will be transfixed by the slightest sound or unexpected occurrence, frozen in mid-movement until the potential enemy has left the scene. This petrified stance can be maintained for hours, so long as the danger lasts, and will frustrate even the most patient watcher! The strategy of cryptic locomotion can only work in luxuriantly leafy surroundings, and, unlike the bush babies, the potto and lorises only inhabit the thickest vegetation.

Should one of these animals be discovered by a small carnivore, certain adaptations enable them to defend themselves and sometimes dissuade the aggressor. The nape and the back of the potto, for example, are protected by a shield of thickened skin overlying hump-like protuberances formed by the spinal processes on the vertebrae projecting through shoulder blades which nearly meet in the middle of the back. The "shield" is covered by fur and by tactile hairs 2–4in (5–10cm) long which detect any attack. In the event of an attack by, for example, an African palm civet, the potto turns toward it, head buried between its hands and presenting its shield. The aggressor's charges are dodged by sideways movements without the potto loosening the grip of its hands and feet on the branch. Then, straightening its body, and maintaining a clamp-like hold on the branch, the potto delivers fearful bites or a violent blow with its shield, toppling the predator to the ground, where it is difficult for the aggressor to find a route back to its prey.

The Golden potto or angwantibo is much more slender than the potto, it has no shield and is incapable of such defense. Faced with a predator it will roll into a ball, completely hiding its head and neck within one arm and its chest. Only the small button-like tail emerges from this motionless, hairy ball, its erectile hairs raised in the form of a ring. The odd posture can puzzle a predator, which may approach carefully and sniff at the tail of the animal. As soon as this happens or it is seized by its rump, the angwantibo directs a bite under its arm at the opponent. The predator's abrupt recoil may throw the angwantibo several meters, where it will once again roll up into a ball.

All members of the family have a mixed diet of small prey, mostly insects, together with fruits and gums. Generally the smaller species feed more on prey (70–80 percent of diet), while the larger species eat more fruit and gums (70–80 percent).

The bush babies as they move rapidly through the foliage disturb many prey animals, whose movements betray them. Their highly developed ears have a series of folds which enable them precisely to orientate the outer ear towards a sound source, much like a bat. Experiments in captivity have shown that a bush baby could follow the movements of a flying or walking insect even through an opaque partition. Many prey items, such as moths and grasshoppers, are taken as they fly past. Keeping its feet clamped fast to the support, the bush baby suddenly extends its body and grabs the prey with one or both hands. These stereotyped hand movements associated with locating the prey by hearing are so precise that a Dwarf bush baby can catch gnats on the wing. Bush babies grasp their prey between the palm and the fingertip of the little finger, the only one with a fleshy pad.

The pottos and lorises, by contrast, have short fingers with soft pads on the tips of all fingers. They catch their prey, which is mostly slow moving and detected by smell, when it is stationary. Typically the prey is foul-smelling or bears hairs that cause irritations; these items are rejected by most predators, particularly bush babies, but pottos and lorises eat them with little problem—irritant caterpillars and butterflies, ants, fleas, foul-smelling beetles and poisonous millipedes. The angwantibo feeds mainly on irritant caterpillars, holding the head between its teeth as it rubs the insect vigorously with both hands to remove some of the hairs before chewing it up. When it has swallowed the prey it wipes its lips and snout clean on a branch. If an angwantibo is given the choice between such caterpillars and more "edible" items such as grasshoppers or hawk moths, it will reject the caterpillars. The ability to eat prey which is not very palatable yet is easy to locate by its scent seems to be a consequence of the slow, imperceptible movement that is characteristic of pottos and lorises.

Compared with other mammals of similar size, members of the family have a relatively low reproduction rate. With few exceptions they reproduce only once a year, usually giving birth to a single young.

In bush babies, the newborn is covered in a fine down, its eyes are half open and it is unable to move about itself. The mother leaves it in the nest briefly only at the beginning, and after 3–4 days carries her offspring by the flank in her mouth. She then parks the young bush baby on a small branch and goes to feed nearby. As she moves from one spot to another during the night she carries her offspring, returning to the nest only at dawn. After 10–15 days, mother and young rejoin the group (several related mothers with young of different ages). In the largest species, the Thick-tailed bush baby and the Needle-clawed bush baby, at about one month the young is able to cling to its mother's back. It subsequently follows its mother (or other individuals of the same group) at gradually increasing distances.

Pottos and lorises do not make a nest and are more developed at birth. The thickly furred newborn clings to the belly of its mother, who carries it there for several days. Very soon, the mother deposits her young on a branch (baby parking) and only retrieves it later in the night when moving to forage elsewhere or even in the morning when going off to sleep. Later, the offspring follows its mother as she moves about, first clinging to her back, then following her over longer and longer distances, learning from her to recognize different types of food. Weaning is at 40 to 60 days, and the young enters puberty between 8 and 12 months of age.

The bush babies have the most complex social structure of the family. The female occupies a territory whose limits she indicates to her neighbors. The growing offspring accompanies nearly her every move

► **Moving about in the trees.** The bush babies are agile leapers with long hindlimbs and bushy tails used for balance when jumping. (1) Thick-tailed bush baby (*Galago crassicaudatus*). (2) Dwarf bush baby (*G. demidovii*). (3) The needle-clawed bush babies (here, *G. elegantulus*) have needle points that help to grip on trees on the nails of all digits except the thumb and big toe.

Lorises and pottos are slow-moving climbers with a strong grip, opposable first digit, and no tail. (4) The Slender loris (*Loris tardigradus*) has a particularly mobile hip joint for climbing. (5) Slow loris (*Nycticebus coucang*). (6) Potto *Perodicticus potto*). (7) Angwantibo or Golden potto (*Arctocebus calabarensis*).

1
4
5
6
7
3

and so occupies the same territory. Males leave the mother's territory after puberty, but a young female maintains the association with her mother. Small social groups are thus formed comprising mothers, daughters and sisters and their young. Females from outside the group are chased off the shared territory, within which there may be areas that are primarily used by one or another group female.

Usually a single adult dominant male mates with the females of a group, but the same male may also control other females outside the group. Male territories are much bigger in area than those of females, and competition between males that are not established is intense. While females remain for years in the same place, dominant males are replaced almost every year. Young adult males that have gone through a period of wandering shortly after puberty establish themselves near the territory of a matriline group, sometimes forming small groups of 2–3 bachelor males awaiting the chance of supplanting a dominant male. Males bear the scars of fights for possession of a female group, and in captivity such struggles may result in death unless the combatants can be separated in time. In some species, such as the Dwarf bush baby, some small adult males remain in the female group, but in such cases they behave towards the females

HBL = head-and-body length; TL = tail length; wt = weight.

Bush babies or galagos
Subfamily Galaginae

Six species, all in Africa. Very large eyes adapted to nocturnal vision; long tail with tuft; large membranous ears which can be folded up; long hind limbs adapted to leaping. Most construct a nest.

Allen's bush baby
Galago alleni

Gabon, N Congo, SE Central African Republic, S Cameroon and S Nigeria. In tropical rain forest. Living chiefly in lower storey, leaping from small trunk to trunk and between bases of lianas. HBL 8in; TL 10in; wt 9oz. Coat quite thick, smoky-gray with reddish flanks, thighs and arms; underside pale gray. Coloration of the tail in Gabon populations may range from dark to silver-gray, with or without 0.5–2in white tip. Diet: fruits, small prey. One, usually single, birth in a year. Gestation: 133 days; newborn weighs 1oz. Female territory about 25 acres. Equal numbers of males and females at birth but later ratio 1:4. Density in Gabon 40/sq mi. Longevity: to 12 years in captivity.

Lesser bush baby
Galago senegalensis
Lesser or Senegal bush baby.

Most widely distributed species, in a vast area bounding African rain forests from Senegal to N Ethiopia, Somalia to Natal, Mozambique to Angola, in dry forests, gallery forests and savanna with trees. HBL 6in; TL 9in; wt 9oz. Smaller than Allen's bush baby, with paler gray coat, longer ears and hind limbs. Diet very varied, including small prey, *Acacia* gum, fruits, nectar. Gums provide the basic food in dry periods. Reproduction twice a year; often twins, exceptionally triplets born. Gestation: 123 days; newborn weighs 0.5oz. Density up to 700/sq mi in some regions. Longevity: to 14 years in captivity. Subspecies 9: *albipes, braccatus, dunni, gallarum, granti, moholi, senegalensis, sotikae, zanzibaricus.*

Thick-tailed bush baby
Galago crassicaudatus

E and S Africa in dense dry and gallery forests, often in same areas as Lesser bush baby. HBL 12in; TL 17in; wt 3lb; the largest bush baby and the one least adept at leaping. Coat gray. Accorded separate genus status by some (*Otolemur crassicaudatus*). Diet: similar to Lesser bush baby but prey items often bigger (birds, eggs, small mammals, reptiles), although not in all subspecies. Although area of distribution only half that of *G. senegalensis*, 10 subspecies distinguished based on size and tint of gray: *argentatus, lönnbergi, umbrosus, garnetti* (sometimes accorded species status), *monteiri, badius, crassicaudatus, lasiotis, agysimbanus, kikuyensis.*

Dwarf bush baby
Galago demidovii
Dwarf or Demidoff's bush baby or galago.

In three distinct areas: C Africa, Gabon, Central African Republic, Uganda, W Tanzania, Burundi, Zaire, Congo; Senegal, S Mali, Upper Volta, SW Nigeria, Dahomey to Senegal; very small area on coast of Kenya. In tropical rain forest, in thick foliage and lianas, crowns of tallest trees, beside tracks, in former plantations. HBL 5in; TL 7in; wt 2oz; smallest lorisid and, with *Microcebus murinus* of Madagscar, smallest of all primates. Coat: gray-black to reddish, depending on the individual and age (darker in young animals). Accorded separate genus status by some (*Galagoides demidovii*). Diet chiefly small insects but also fruits and gums. In Gabon one, usually single, young per year (1 in 5 births twins); gestation 110 days; newborn weighs 0.2–4oz, reaching adult size at 2–3 months and puberty at 8–9 months. Female territories about 2.5 acres, dominant males' about 4 acres. Density about 130/sq mi. Longevity: to 12 or more years in captivity. Subspecies: 7 based on variation in size and coat color, but some doubtful as variations may occur in some populations: *animurus, demidovii, murinus, orinus, phasmus, poenis, thomasi.*

Needle-clawed bush baby
Galago elegantulus

Gabon, S Nigeria, S Cameroon, N Congo. In forests. HBL 12in; TL 12in; wt 10oz. Coat: an attractive reddish color on back, with darker line down back, ash-gray underside

▲ **Just half an ounce in weight,** a newborn Lesser bush baby clings in adult posture to the stalk where it has been placed.

◄ **Shining eyes** of this angwantibo, or Golden potto, reflect the photographer's flash. Night vision is aided by the *tapetum*, a layer of cells behind the retina of the eye which reflects the light back through the retina – hence the "eyeshine."

and dominant male as would an immature individual.

The dominant male regularly visits all his females in order to monitor their sexual cycle. When a female comes into heat, the male will not leave her until they have mated.

At night-time bush babies communicate by loud cries and they mark with urine throughout their travels by an unusual method; this "urine washing" involves balancing on one foot, depositing drops of urine in the hollow of the other foot and of the hand on the same side, subsequently rubbing one against the other. The same operation is carried out standing on the other foot. In this way the animal's path is marked with a scent of urine that has important social functions. By means of scent and sound, bush babies can maintain social relations at a distance ("deferred communication"). Not until morning do the bush babies gather again, using a special rallying call, before going to sleep as a group in a hole in a tree, in a clump of branches, or in a round nest made of green leaves piled up in the fork of a tree.

Among pottos and lorises there is a simpler social structure based on the same territorial system. There are no matrilinear groups of females, and social communication is principally by means of the messages contained in urine marking.

Members of the family feature in certain myths of African societies. In the ethnic groups of Gabon, for example, every species has its name. Allen's bush baby is called "Ngok" or "Nogkoué" in onomatopoeic style after the species' alarm call. It is said to warn of the approach of a leopard, a belief not far from the truth since Allen's bush baby lives in the undergrowth and will at once call out in alarm should a "suspicious" animal pass nearby.

A more obscure tale concerns the potto, which is supposed, if caught in a trap, to be able to hold an antelope prisoner in its vice-like grip until the hunter returns to his snare. This widespread myth celebrates the great strength of the potto, one of the largest members of the family.

Because of their small size and nocturnal habits, lorisids are not hunted systematically like monkeys and other vertebrates in the countries where they live. In fact, few people have seen these animals in the wild and apart from zones where their habitat is threatened with destruction, man currently presents little real danger to their survival. Sometimes a loris, bush baby or potto is brought to Europe as a pet, but the temperature of an apartment and the lack of insects sooner or later prove fatal to them.

PC-D

and flanks; tail gray always with white tip. Often placed in separate genus *Euoticus* on basis of differences indicated below. Nails elongated to form fine point; branches etc may be held by pad on last joint of digits as in other bush babies, or by digging in nails. Diet 80% gums with corresponding adaptations: second premolar in form of canine, elongated dental comb, longer intestine. The heavier muzzle than other bush babies, and large golden eyes give it a striking appearance. The only bush baby not to make a nest or take refuge in holes in trees. Reproductive biology not studied. Longevity: to 15 or more years in captivity. Subspecies 2: *G. e. pallidus* in North and *G. e. elegantulus* in South.

Needle-clawed bush baby
Galago inustus

In Great Rift Valley in Rwanda, Burundi, E Uganda. HBL 6in; TL 9in; wt 9oz. Coat: very dark. Separated from Lesser bush baby on basis of coat color, sometimes placed in genus *Euoticus* on basis of nails which are similar to those of *G. e. elegantulus*. Biology unknown.

Lorises and Pottos
Subfamily Lorisinae

Two species in Africa, 2 in Asia. Tail very short, limbs less unequal in length than in bush babies, adapted for climbing. Hands and feet developed into pincers with opposable thumb and much reduced index finger. Do not construct a nest.

Potto
Perodicticus potto

In tropical forests of W African coast from Guinea to Congo and from Gabon to W Kenya. HBL 12in; TL 2in; wt 2lb. Large, muscular, compact. Coat: reddish-brown to blackish; ears often yellowish within. Processes project from vertebrae between shoulder blades to help form "shield" on nape and back. Diet: chiefly fruits, some gums, small, often irritant prey items (birds, bats, rodents) eaten whole. After 193-day gestation, a single young born dark-colored in Gabon, speckled with white in Ivory Coast, weighing 2oz. Young attain puberty at about 1 year. Home range area of females about 18 acres, of males about 30 acres. Density about 20/sq mi. Longevity: to 15 or more years in captivity. Subspecies 5: *edwarsi, faustus, ibeanus, juju, potto.*

Angwantibo
Arctocebus calabarensis
Angwantibo or Golden potto.

S Nigeria, S Cameroon, Gabon, Congo, W Zaire. In Gabon, in wet, low forest rich in lianas, also in former scrubby plantations (where it preys chiefly on caterpillars). HBL 9in; TL 0.5in; wt 7–18oz. Slender and light compared to the heavier compact *P. potto*. Coat an attractive light reddish color. Gestation 135 days; one young born weighing 1oz with white-spotted coat. Mother mates a few days after birth, so weaning occurs a few days before the next birth. Subspecies 2: *calabarensis* in north, 14–18oz; more slender and smaller *aureus* in south, 7oz.

Slender loris
Loris tardigradus

India and Sri Lanka. HBL 9in; no tail; wt 10oz. Coat gray or reddish, varies according to subspecies. Similar slender build to angwantibo but larger. Eyes surrounded by two black spots separated by narrow white line down to nose. Biology similar to angwantibo's. Subspecies 6: *grandis, lydekkerianus, malabaricus, nordicus, nycticeboides, tardigradus.*

Slow loris
Nycticebus coucang

Bangladesh to Vietnam, Malaysia, Sumatra, Java, Borneo. HBL 12in; TL 2in; wt 3lb. Anatomy and life-style similar to potto, coat color more varied: ash-gray, darker dorsal line divides on head into two branches which surround eyes. One infant, 2oz. Subspecies 10: *bancanus, bengalensis, borneanus, coucang, hilleri, insularis, natunae, javanicus, tenasserimensis, pygmaeus* (this last, in Indochina, is recognized as a separate species by some authorities).

TARSIERS

Family Tarsiidae
Three species of genus *Tarsius*.
Distribution: islands of SE Asia.

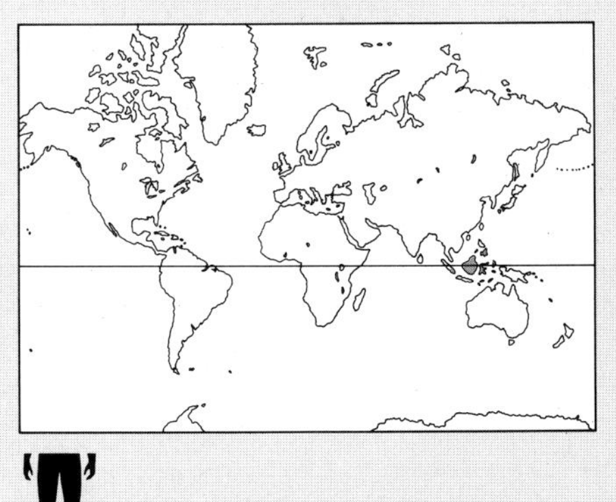

Philippine tarsier [E]

Tarsius syrichta
Distribution: SE Philippine Islands (Samar, Mindanao etc).
Habitat: rain forest and shrub; crepuscular and nocturnal.
Size: head-body length 4.3–5in (11–12.7cm); tail 8.3–10in (21–25cm); weight 3.9–4.2oz (110–120g).
Coat: gray to gray-buff; face more ocher; tail tuft sparse.
Longevity: in captivity 8–12 years.

Western tarsier

Tarsius bancanus
Distribution: Borneo, Bangka, S Sumatra.
Habitat: primary and secondary rain forest, shrubs, plantations; lives in pairs; crepuscular and nocturnal; insectivorous and carnivorous.
Size: head-body length 4.5–5.7in (11.5–14.5cm); tail 7.9–9.2in (20–23.5cm); weight 3.7–4.8oz (105–135g).
Coat: buff, brown-tipped; tail tuft well developed but not bushy.
Longevity: 8 years or more.

Spectral tarsier

Tarsius spectrum
Spectral tarsier, Celebes or Sulawesi tarsier.
Distribution: Sulawesi, Great Sangihe, Peleng.
Habitat: primary and secondary rain forest, shrubs, plantations; lives in pairs or family groups; crepuscular and nocturnal; insectivorous.
Size: head-body length 3.7–5.5in (9.5–14cm); tail 7.9–10.2in (20–26cm); weight probably slightly less than either of the other species.
Coat: gray to gray-buff, sometimes darker than other species; tail tuft long and bushy, scale-like skin at tail.
A small, montane subspecies (*T. s. pumilus*) may be separated as the species *Tarsius pumilus*.

[E] Endangered.

For the head-hunting Iban people of Borneo the tarsiers once played an important role as a totem animal, since the head of this small, nocturnal primate was believed to be loose. This belief stemmed from the extraordinary capability for rotation of the tarsier's neck vertebrae. The crucial systematic position of the tarsiers, between other prosimians and monkeys, makes them relevant to many problems of primate evolution.

The three species are of similar size. Tarsiers are buff-gray or ocher, sometimes beige or sand-colored. The coat is softer than velvet. These "vertical clingers and leapers" are famous for their leaping abilities. Head-and-body length is only about half as long as the whole hind limb. All three segments of the leg (thigh, lower leg, and foot) are elongated and roughly equal in length (in the Western tarsier 2–3in each).

In the Spectral tarsier the long slender tail is covered in "scales" like those of mice and rats, but in the other two species only the surface of the skin shows how such scales were once arranged. In each species the tuft of hairs on the tail has a distinctive form. The second and third toes of tarsiers are equipped with a so-called toilet claw used for grooming, while the other toes bear nails, as do the fingers. The fingers are long and slender, and are used as a kind of cage to trap swift insect prey in the darkness of a forest night. The third finger of the Western tarsier is roughly the same length as the upper arm (about 1.2in/3cm).

The enormous size of the eyes indicates the tarsiers' nocturnal predatory habits. They are directed forward to allow stereoscopic vision and, like those of owls, can hardly be moved within their orbit. In a Western tarsier each eye weighs slightly more than the whole brain (about 0.1oz) and in the brain the visual regions predominate. Tarsiers first locate many of their prey with their sharp ears. In comparison to other primates of their size the tarsiers have needle sharp, rather large teeth.

While fossil relatives have been found in Asia, Europe, and America, the tarsiers of today are restricted to some southeast Asian islands. All three species occur separately. The Spectral tarsier from Sulawesi is the most primitive of modern tarsiers and the one least specialized to both nocturnal activity and to exclusively leaping movement between vertical trunks and branches. Sulawesi is separated from the Philippines and Borneo by Wallace's Line, which marks the division between the Eurasian and the Australasian fauna. *Tarsius* is unusual in that its distribution crosses this zoogeographic border. This is an indication of the tarsiers' long residence in the region, which may go back more than 40 million years. Modern tarsiers may derive from the ancestor of the more primitive Sulawesi species. The Philippine and Western tarsiers share a number of characters, which cannot be found in the Sulawesi tarsiers.

All three species seem to be exclusively insectivorous and carnivorous. All kinds of arthropods are taken, including ants, beetles, cockroaches or scorpions. Variation in diet is great; different individuals may relish or disregard lizards and bats. A Western tarsier can catch and kill a bird larger than itself. Venomous snakes are also sometimes taken. Tarsiers also drink several times each night. Prey is caught invariably by leaping at it, pinning it down by one or both hands, and killing or at least immobilizing it by several bites. The victim, often caught on the ground, is carried in the mouth to a perch, where it is eaten head first. A Western tarsier may eat about 10 percent of its own weight per day (0.35–0.5oz/10–14g).

Western tarsiers are sexually mature at about one year. In this species and the Sulawesi tarsier births occur throughout the year, although Western tarsier births peak at the end of the rainy season (February–April).

Courtship in the Western tarsier is accompanied by much chasing around, sometimes with soft vocalizations. During mating, which occurs when sitting in a tree, both partners are silent. The gestation period is about six months. The single young is born fully furred and with its eyes open. It can climb around in its first day of life and at birth weighs almost one quarter as much as its mother. Before hunting, the mother will "park" her offspring, who can call her with soft clicking or sharp whistling calls, depending on how far away she is.

In the Western tarsier, one pair per home range, possibly with one young, seems to be the rule. One Sulawesi group has been observed to occupy about 2.5 acres (1ha), giving a density, in a favorable area, of about 650–900 tarsiers per square mile (250–350/sq km).The home ranges overlap to some extent, but core areas seem to be defended, as indicated by frequent injuries (including typical fractures of bitten fingers). However, all three species reduce the risk of open fights by marking with urine and with a secretion from a skin gland on the chest (epigastric gland). In the Sulawesi species a pair or small family group may stay together, uttering social calls near the sleeping site in the center of the home range. In these spacing calls the male and female perform beautiful duets with very different, high-pitched voices.

On Sulawesi and on Borneo tarsiers are abundant and not endangered, except where the forest is logged—logging certainly kills tens of thousands every year. In Malaysia and Indonesia tarsiers are protected by law. It seems that their only true sanctuaries will be national parks.

The appearance of tarsiers appeals to many people, who sometimes try to keep them as pets. But as these delicate primates require appropriate live food, they usually die within days. All tarsiers examined have had intestinal worms (hookworms, common ascarid worms, tapeworms and others). Man may be susceptible to some of these.

CN

▲ **The three species of tarsier,** showing the extraordinary proportions of the hindlimbs, the toilet claws used for grooming, and differences in tail-tuft patterns. (1) Spectral tarsier (*Tarsius spectrum*) or Celebes tarsier. (2) Philippine tarsier (*T. syrichta*). (3) Western tarsier (*T. bancanus*).

▼ **Vertical clinger and leaper.** The forward-seeing eyes allow the tarsier to judge its safe landing, and each eye may equal the brain in weight.

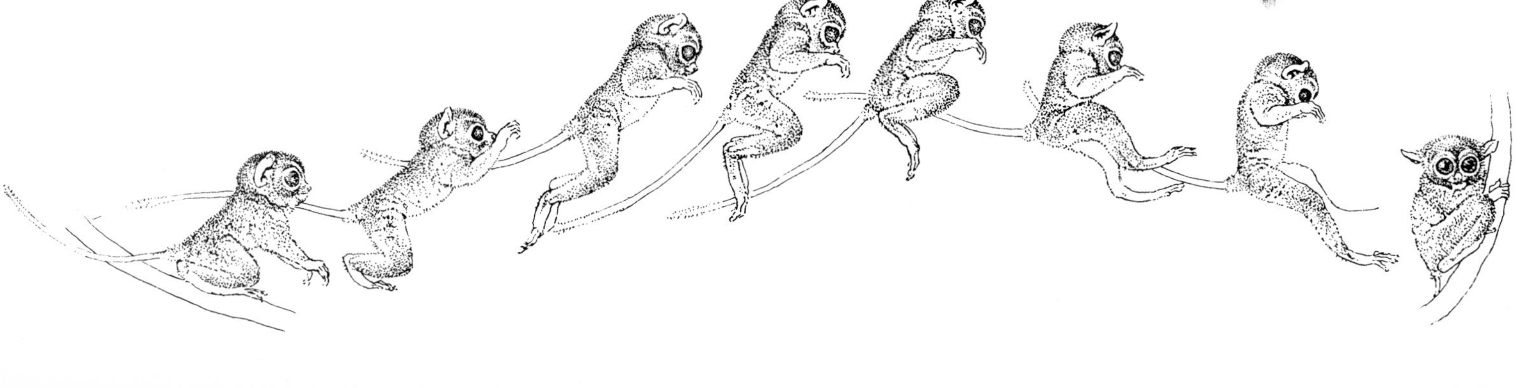

MONKEYS

Three families: Cercopithecidae, Cebidae, Callitrichidae.
One hundred and thirty-three species in 30 genera.
Distribution: chiefly within the tropics, in S and C America, Africa and Asia.

Habitat: chiefly forests, some species in grasslands.

Size: ranges from the Pygmy marmoset, with head-and-body length 7–7.5in (17.5–19cm), tail length 7.5in (19cm) and weight 4.2–6.7oz (120–190g), to the drill and mandrill, males of which may measure 12in overall (31cm), and weigh up to 110lb (50kg).

Marmosets and tamarins (family Callitrichidae)
Twenty-one species in 5 genera.

Capuchin-like monkeys (family Cebidae)
Includes **Howler**, **Woolly** and **Spider monkeys**, **Night** and **Squirrel monkeys**, **titis sakis**, **capuchins**, and **uakaris**.
Thirty species in 11 genera.

Old World monkeys (family Cercopithecidae)
Guenons, macaques and baboons (subfamily Cercopithecinae), also **mandrills** and **mangabeys**.
Forty-five species in 8 genera.

Leaf monkeys (subfamily Colobinae), including **colobus monkeys** and **langurs**.
Thirty-seven species in 6 genera.

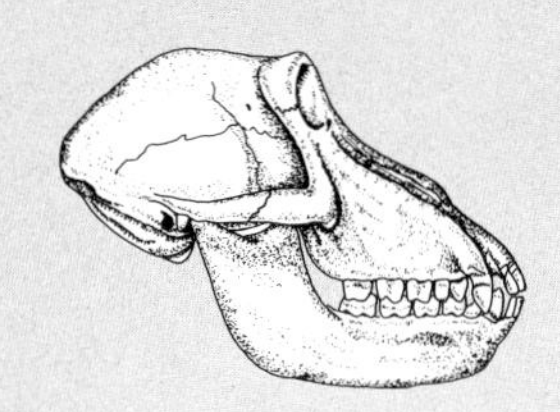

Savanna baboon (female) 6in

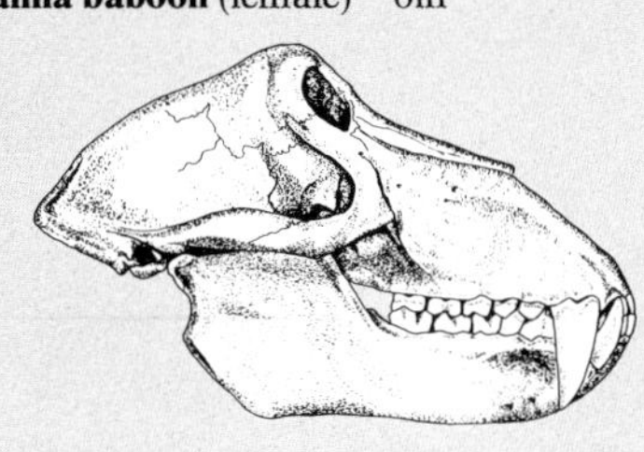

Savanna baboon (male) 8in

The monkeys differ from the prosimians in the detailed structure of their skulls, the structure of their placenta and the relative size of their brains. In general, they show an increased reliance on sight over sound and smell and greater flexibility in their mode of locomotion.

The first monkeys appeared in the Oligocene, around 35 million years ago. Present-day monkeys are divided into two geographically separate lineages—the New World monkeys, sometimes called the platyrrhines, and the Old World monkeys, or catarrhines. These names are derived from the shape of the nose, which provides a reliable way of distinguishing between the two groups: New World monkeys have nostrils that are wide open and far apart (platyrrhine) while in Old World species the nostrils are narrow and close together (catarrhine). Other obvious differences include the development of a prehensile tail in the larger-bodied species of New World monkeys but not in Old World monkeys, and the development of ischial callosities—hard, sitting pads on the lower side of the buttocks—in Old but not New World monkeys.

It used to be thought that the New World and Old World monkeys had developed separately from Eocene prosimians and that their similarities were the result of convergent adaptations to a common way of life. However, more recent biochemical evidence links the two groups more closely, clearly distinguishing them from the prosimians.

The New World monkeys consist of two main groups. The marmosets and tamarins are all small-sized fruit-eaters mostly weighing around 0.6lb (300g). All the 21 species

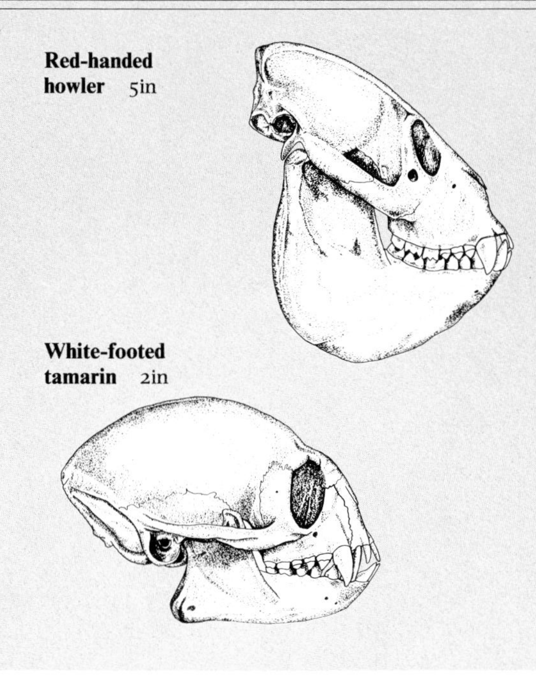

Skulls of Monkeys

In all monkeys, the eyes are directed forwards for binocular vision and are contained in bony sockets produced by a virtually complete plate behind the orbit. The frontal bones of the forehead become fused together early in life. The brain-case is quite large and globular, reflecting the relatively large brain size of the monkeys. The two halves of the lower jaw are fused together at the mid-line and the body of the jaw is typically fairly deep.

Monkeys have spatulate (shovel-shaped) incisors, conspicuous canines, and squared-off molar teeth which typically have four cusps. In marmosets, the lower incisors are as tall as the canines (see box p48). Among the New World monkeys, the capuchin-like monkeys (family Cebidae) all have a formula of I2/2, C1/1, P3/3, M3/3 = 36, while both marmosets and tamarins (family Callitrichidae) have I2/2, C1/1, P3/3, M2/2 = 32 and are the only primates to have reduced the dental formula by loss of molar teeth. As in other respects (see p46), Goeldi's monkey (*Callimico*) is intermediate in that it has the full dental formula of the capuchin-like monkeys but has tiny third molars in both upper and lower jaws. In Old World monkeys there is but a single dental formula (shared with apes and man) which has been attained by losing premolars: I2/2, C1/1, P2/2, M3/3 = 32. The molars of Old World monkeys are distinctive in that the four cusps are joined in pairs by transverse ridges (bilophodonty, see p12).

Sexual dimorphism, often marked in monkeys, is less pronounced in New World species, among which howler monkeys are the most extreme case: their tendency towards leaf-eating is reflected in the depth of the lower jaw. In the Old World baboons the male can be twice as heavy as the female and the skull is accordingly much larger.

of this family are active in the daytime and arboreal, and the majority live in small groups consisting of a monogamous breeding pair and their offspring. Among monkeys, they are unique in that all their digits, except the first toe which bears a flat nail, end in a sharp, curved claw rather than a flat nail, which helps them to run up the sheer sides of the big trees in the South American rain forests where they live. They are strongly territorial and have shrill, bird-like calls.

In contrast, the capuchin-like monkeys (sometimes, but not here, termed "the New World monkeys") are a much more diverse family. They consist of four principal groups, sometimes classified as subfamilies, which occupy contrasting ecological niches. The howler monkeys are large, leaf-eating animals weighing 13–18lb (6–8kg) which digest their coarse diet with the help of a special extended cecum. They live in groups of 2–30, sometimes including several adult males. The spider monkeys and woolly monkeys are similar in size to howlers but feed on a mixed diet of fruit and leaf shoots and range widely. They have larger brains, relative to body size, than most of the other New World monkeys and very long periods of infant dependence—of all the South American primates they most closely resemble the African apes. They have very large ranges and can travel several miles in a day. The sakis and uakaris are smaller in size and specialize in feeding on the seeds of forest trees. The last group includes the day-active fruit-eating capuchins, the dainty squirrel monkeys and monogamous titis, and the nocturnal Night monkey.

The radiation of Old World monkeys is probably the most recent among the major primate groups. Like the New World monkeys, the Old World monkeys may be grouped into subfamilies. The larger of the two subfamilies, the Cercopithecinae, is sometimes known as "typical monkeys." It includes the African baboons and the Asian macaques which are mostly large (22–44lb/10–20kg), omnivorous and terrestrial and live in open country. The baboons are replaced in the African rain forests by the terrestrial mandrills and the smaller, arboreal mangabeys. The drills and mandrills live in large troops which vary in size throughout the year and are thought to consist of harem units as in Hamadryas baboons, while the mangabeys occur in stable groups of 10–30 animals. The bare highlands and deserts of Ethiopia are occupied by the Hamadryas baboon and the gelada. In Asia, the macaques replace the baboons and include both arboreal and terrestrial species which are found in forested country as well as in more open areas. One species of macaque extends into North Africa and Gibraltar, where it has been (mis)named the Barbary ape.

The other group of Old World monkeys are the colobine monkeys—which includes nine species of African colobus monkeys and 26 Asian leaf monkeys. The distinctive feature of all these animals is that, unlike any other primates, they all possess a large forestomach which contains populations of microbes that are able to digest the cellulose in the foliage that the animals eat. The colobines are the primate equivalent of ruminants and are specialized folivores, though they will also eat fruit and flowers when these are available. THC-B

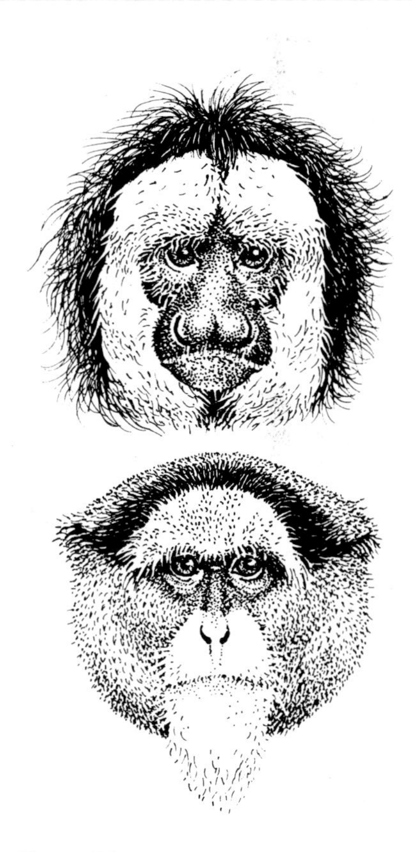

▲ **Shape of the nose** distinguishes New World from Old World monkeys. The White-faced or Guianan saki (above) from north of the Amazon has nostrils that are wide open, far apart and face outward; it is platyrrhine (broad-nosed). In De Brazza's monkey, a guenon from African swamp forests, the nostrils are narrow, close together and point downward (catarrhine = downward-nosed).

◄ **Vivid colors** on the bare face of an adult male mandrill, largest of monkeys. The bright red pigment depends on gonadal hormones, so is absent in juveniles. The blue requires similar hormone but once established becomes structural and permanent.

MARMOSETS AND TAMARINS

Family: Callitrichidae.
Twenty-one species in 5 genera.
Distribution: S Central and N half of South America.

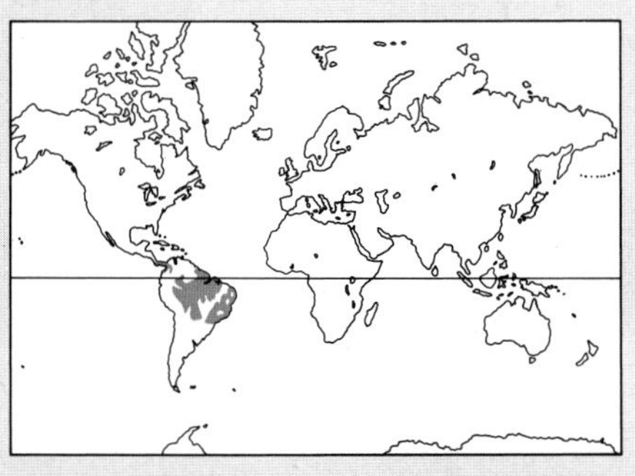

Habitat: chiefly tropical rain forest, also gallery forest and forest patches in savanna.

Size: ranges from Pygmy marmoset, head-body length 7–7.5in (17.5–19cm), tail 7.5in (19cm), weight 4.2–6.7oz (120–190kg), the smallest of all monkeys, to the Lion tamarin, head-body length 13–16in (34–40cm), tail 10–15in (26–38cm), weight 22.2–25oz (630–710g). Most species weigh 9.2–13.4oz (260–380g).

Coat: fine, silky, often colorful. Many species have ear tufts, mustaches, manes or crests.

Gestation: 130–170 days.

Longevity: not known in wild (to 7–16 years in captivity).

Species include:
Pygmy marmoset (*Cebuella pygmaea*).

Marmosets (*Callithrix*, 7 species), including the **Tassel-ear marmoset** [V] (*C. humeralifer*), **Bare-ear marmoset,** Silvery, or Black-tailed marmoset (*C. argentata*), **Common marmoset** (*C. jacchus*) and **Black tufted-ear marmoset** or Black-eared marmoset (*C. penicillata*). **Buffy-headed marmoset** [E] (*C. flaviceps*)

Tamarins (*Saguinus*, 11 species), including the **Emperor tarmarin** [I] (*S. imperator*), **Red-handed tamarin** or Midas tamarin (*S. midas*), **Cotton-top tamarin** (*S. oedipus*), **Silvery-brown bare-face tamarin** [V] or White-footed tamarin (*S. leucopus*), **Geoffroy's tamarin** (*S. geoffroyi*), **Moustached tamarins** (*S. mystax*, *S. labiatus*), **Saddle-back tamarin** (*S. fuscicollis*), and **Pied tamarin** [I] (*S. bicolor*).

Lion tamarin [E] (*Leontopithecus rosalia*).

Goeldi's monkey [R] (*Callimico goeldii*).

[E] Endangered. [V] Vulnerable. [R] Rare.
[I] Threatened, but status indeterminate.

With their fine, silky coats, long tails and a wide array of tufts, manes, crests, mustaches and fringes, marmosets and tamarins are the most diverse and colorful of the New World primates. These diminutive, squirrel-like monkeys of the tropical American forests share a combination of features that is extremely unusual among primates, including a monogamous breeding system, with few differences in form between the sexes, multiple birth (usually twins), extensive care of young by the father, and social groups as large as 15 comprised of the breeding pair and their offspring. The latter may remain in the group, even when adult, without breeding but help to care for their younger siblings. Marmosets (but not tamarins) are uniquely specialized gum eaters (see p48).

With one exception, marmosets and tamarins are distinguished from the other monkeys of the New World, the capuchin-like family Cebidae, by their small size, modified claws rather than nails on all digits except the big toe, the presence of two as opposed to three molar teeth in either side of each jaw, and by the occurrence of twin births. These features, together with their simple uterus and their lack of a rear inner cusp on the upper molars, have led to the suggestion that marmosets and tamarins are advanced primates which have again become small during their evolution, as an adaptation to the adoption of an insectivorous diet. Goeldi's monkey—a tamarin—is also believed to have undergone phyletic dwarfism but shares traits with both the Cebidae (three molar teeth in each jaw and single offspring) and the Callitrichidae (small size and claws rather than nails).

The Pygmy marmoset and Goeldi's monkey are the only representatives of their respective genera and both are restricted to the upper Amazon, in Brazil, Peru, southern Colombia and northern Bolivia. Goeldi's monkeys prefer to forage and travel in the dense scrubby undergrowth and, possibly because of this habitat preference, they are rare, with groups in patches of suitable vegetation isolated from one another by several miles. The Pygmy marmoset is also a habitat specialist, and populations are at their highest—up to 104–130 groups/sq mi (40–50 groups/sq km)—in the riverside and seasonally flooded forest where their preferred tree gum sources are most abundant. In areas away from rivers and in secondary forest they occur in lower densities of 26–31 groups/sq mi (10–12 groups/sq km).

The Tassel-ear and Bare-ear marmosets

▼ ► **Marmosets and tamarins:** variety in form and color of facial skin and head hair, foraging behavior, scent marking and offensive threat postures. (1) Goeldi's monkey (*Callimico goeldii*) in "arch-bristle" offensive threat posture used within the troop. (2) Pygmy marmoset (*Cebuella pygmaea*) gouging tree for gum and sap. (3) Silvery marmoset (*Callithrix argentata argentata*) and (4) Black-tailed marmoset (*C. a. melanura*), the latter presenting its rear with tail raised as an offensive threat posture used between members of different troops. (5) Buffy tufted-ear marmoset (*C. aurita*). (6) Black tufted-ear marmoset (*C. penicillata*). (7) Tassel-ear marmoset (*C. humeralifer intermedius*) eating typical fruit item. (8) Golden-rumped Lion tamarin (*Leontopithecus rosalia chrysomelas*) using elongated fingers to probe a bromeliad for insects. (9) Geoffroy's tamarin (*Saguinus geoffroyi*) scent marking a branch with glands situated around its genitals. (10) Red-chested moustached tamarin (*S. labiatus*) marking with chest glands. (11) Saddle-back tamarin (*S. fuscicollis*) marking with glands above pubic area. (12) Mottle-faced tamarin (*S. inustus*). (13) Emperor tamarin (*S. imperator*). (14) Black-chested moustached tamarin (*S. mystax*). (15) Cotton-top tamarin (*S. oedipus*).

1
5
6
3
9
4
7
8
2
10
12
11

are restricted to Amazonia (except for the dark-coated subspecies of the latter which extends south into east Bolivia and northern Paraguay). The remaining five marmosets occur in southern, central and eastern Brazil. The tamarins, *Saguinus*, are distributed widely through Amazonia, north of the Rio Amazonas and, in the south, west of the Rio Madeira; one subspecies of the Midas tamarin (*S. midas niger*) occurs south of the Amazonas at its mouth. Only three species, the Cotton-top, Geoffroy's, and Silvery-brown bare-face tamarins, occur outside Amazonia, in north Colombia and Central America. These marmosets and tamarins inhabit a wide variety of forest types, from tall primary rain forest with secondary growth patches to semi-deciduous dry forests, gallery forests and forest patches in savanna regions in Amazonia, in the *chaco* of Bolivia and Paraguay and in the *cerrado* of central Brazil. They appear to be most numerous—8–13 groups/sq mi (3–5 groups/sq km)—in areas of primary forest with extensive patches of secondary growth forest, where they feed on fruits of colonizing trees. Bushy vegetation and dense liana tangles of the secondary growth are also their preferred sleeping sites, provide protection from predators such as the marten-like tayra and forest hawks, and probably contain higher densities of their insect foods.

The three subspecies of Lion tamarin survive in widely separated remnant forests in southern Brazil, in low densities of 1–3 groups/sq mi (0.5–1 group/sq km). Although they exploit secondary growth forest, they depend on tall primary forest for sleeping holes in tall tree trunks and for their relatively large animal prey.

Marmosets and tamarins eat fruits, flowers, nectar, plant exudates (gums, saps, latex) and animal prey (including frogs, snails, lizards, spiders and insects). They are generally not leaf-eaters although they infrequently eat leaf buds. The fruits are usually small and sweet, and the genera *Pourouma*, *Ficus*, *Cecropia*, *Inga* and *Miconia* are particularly important. They spend a large portion of their activity time (some 25–30 percent) foraging for animal prey, searching through clumps of dead leaves, amongst fresh leaves, along branches and peering and reaching into holes and crevices in branches and tree trunks. Marmosets also exploit the insects disturbed by Army ant swarm raids. The Pygmy marmoset is primarily an exudate feeder, spending up to 67 percent of its feeding time tree-gouging for gums and saps. Spiders and insects are also important, but fruit is eaten only infrequently. Other marmosets are primarily fruit-eaters, with flowers, animal prey and, particularly at times of fruit shortage, exudates also being important. The remaining members of the family are also fruit-eaters, supplementing their diet with animal prey, flowers and, particularly at times of fruit shortage, nectar. Gums are eaten occasionally when readily available. No marmoset species share the same forest, possibly due to their shared specialization on plant exudates, but a number of tamarin species do (see pp52–53), differing most importantly in the levels and sites at which they habitually travel and forage for animal prey as well as

Gum-eating Monkeys

The Pygmy marmoset and the seven larger marmoset species are uniquely specialized gum eaters. Saps and gums are exuded where the tree is damaged (often from attack by wood-boring insects). The Fork-crowned lemur (p31) uses its tooth-comb to scrape up gums already exuded but marmosets are the only primates which regularly gouge holes irrespective of insect attack. The amount of gum exuded at any such site is usually very small, and they spend only 1–2 minutes licking droplets and gnawing at any one hole. Unlike the tamarins (**a**), the marmosets (**b**) have relatively large incisors, beyond which the canines barely protrude. The marmosets' lower incisors lack enamel on the inner surface, while on the outer surface the enamel is thickened, producing a chisel-like structure. Marmosets anchor the upper incisors in the bark and gouge upwards with these chisel-like lower incisors. The holes they produce are

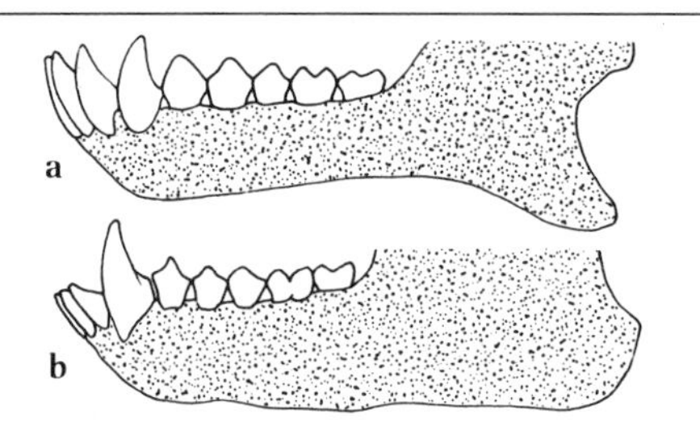

usually oval and about 1in (2–3cm) across at most, but favored trees may be riddled with larger holes and channels as long as 4–6in (10–15cm). The small size and claw-like nails of callitrichids are important adaptations that enable them to cling to vertical trunks and branches. Exudates are a larger part of the diet of the Pygmy marmoset than other marmosets. However, it is an important supplement to their diet, particularly at times of fruit shortage and for such species as the Common and the Black tufted-ear marmosets it is probably vital, allowing them to survive in the relatively harsh environments of northeast and central Brazil.

► **Fruit is a major part of the diet** of the white-faced Saddle-back tamarin of Amazonia. As in all marmosets and tamarins, claws rather than nails are borne by all digits except the big toe.

▼ **Endangered species** of southeastern Brazil, the Buffy-headed marmoset survives only in Atlantic forest remnants.

in their foraging methods. The marmosets and probably the Moustached and Emperor tamarins employ a foliage-gleaning and visual-searching method. The Lion and Saddle-back tamarins are more manipulative foragers, searching in holes, breaking open humus masses, rotten wood and bark, and they possibly exploit larger insects. Lion tamarins have long hands and fingers compared to other callitrichids, an adaptation for this mode of foraging.

Marmosets and tamarins live in extended family groups of between four and 15 individuals. The groups defend home ranges of 0.2–0.7 acres (0.1–0.3ha) in the Pygmy marmoset to 25–100 acres (10–40ha) in the case of *Saguinus* and *Callithrix*, the size depending on availability and distribution of foods and second-growth patches. They visit approximately one-third of their range each day, travelling up to 0.6–1.2mi (1–2km). Range defense involves extended bouts of calling, chasing and, when the groups confront each other, displaying. Marmosets, with their distinct white genitalia, display their rumps with their tail raised and fur fluffed. Lion tamarins raise their manes and tamarins fluff their fur and tongue-flick. Scent marking is also important as a means of registering the group's presence in an area. Marmosets use chest and suprapubic glands for such marking. A third type of scent marking, using the circumgenital glands, is used in communication between group members. This frequently takes place after tree-gouging, when marmosets mark and sometimes urinate in the holes they gouge, possibly as a means of maintaining dominance relations. The most obvious and prolonged social behavior within a group is mutual grooming, most frequently between the adult breeding pair and their latest offspring. Other forms of communication involve a limited number of facial expressions, specific postures and patterns of hair-erection, and complex, graded, high-pitched and bird-like vocalizations.

In the wild only one female per group breeds during a particular breeding season, although she may copulate with more than one adult male in the group. Reproduction is suppressed in other female members of the group, possibly as a result of subordination by the reproductive female as well as chemicals (pheromones) in the scent marks from her circumgenital glands. In Goeldi's monkey groups, there may be two breeding females in a group, although the social organization is otherwise apparently similar; breeding females produce a single young twice a year. In all other species, non-identical twins are born twice a year, at intervals only a few days longer than the gestation period. The birth weights of Pygmy and other marmoset litters are high, 19–25 percent of the mother's weight, which is considerably greater than in other primates. The male parent and other group members help in the carrying of young from birth. Marmoset infants are completely dependent for the first two weeks but by two months can travel independently, catch insects or rob them from other group members, and spend extended periods in play, wrestling, cuffing and chasing each other and other group members. They reach puberty at 12–18 months and adult size at two years of age. In the tamarins, Lion tamarin and Goeldi's monkey, whose newborn are smaller (about 9–15 percent of the mother's weight), males help carry the young only when they are 7–10 days old. Tamarins mature slightly slower than marmosets, becoming independent at 2½ months.

All group members take some part in carrying the young and also surrender food morsels, particularly insects, to them and to the breeding female and infant carriers. So far as is known, this form of cooperative breeding is unique among primates. The adult "helpers" which stay in the family group may gain breeding experience through helping to raise their younger kin, while waiting until suitable habitat becomes available for them to breed or the possibility arises for them to breed in their own or a neighboring group. Once established as breeders, callitrichid monkeys have a higher reproductive potential than any other primate. In suitable conditions a female marmoset can produce twins about every five months.

The variety of species is greatest in the upper Amazon region, where they occur in limited distributions between even quite small river systems. They are therefore extremely vulnerable to extensive habitat destruction. Marmosets and tamarins used to be wrongly considered carriers of yellow fever and malaria and were persecuted in consequence; until 1970–1973 they were captured and exported in large numbers particularly for zoological gardens and for biomedical research. Today only Bolivia, Panama and French Guiana still permit their export. The Lion tamarins (see p54), the marmosets of the south and southeast of Brazil and such forms as the Pied tamarin, which has a minute distribution around the capital of Amazonas in Brazil, face extinction in the near future if habitat destruction continues unchecked. ABR

THE 21 SPECIES OF MARMOSETS AND TAMARINS

Abbreviations: HBL = head-and-body length; TL = tail length; wt = weight.
[E] Endangered. [V] Vulnerable. [R] Rare. [I] Threatened, but status indeterminate.

Marmosets

Genus *Callithrix*

In primary tropical rain forest mixed with second growth, gallery forest, forest patches in Amazonian-type savanna, Bolivian and Paraguayan *chaco*, NE Brazil and *cerrado* of C Brazil. Diet: fruit, flowers, plant exudates (gums, saps, latex), nectar, insects, spiders, frogs, snails and lizards. HBL 7–8in; TL 10–11in; wt 10–12oz. Coat: very variable, some species showing subspecific variation from dark to completely pale or white; varying degrees of ear-tufts. Gestation: 148 days. Litter: 2.

Tassel-ear marmoset [V]

Callithrix humeralifer

Brazilian Amazon between Rios Madeira and Tapajos. Three subspecies: *C. h. humeralifer*—pigmented face, long silvery ear-tufts, mantle mixed silvery and black, back with pale spots and streaks, tail silvery ringed with black, white hip patches; *C. h. intermedius*—face pink, reduced ear-tufts, upper chest and back creamy white, rump and base of tail dark brown, underparts orange; *C. h. chrysoleuca*—face pink, fur near-white, long white ear-tufts, rest of body pale golden to orange.

Bare-ear marmoset

Callithrix argentata
Bare-ear, Silvery or Black-tailed marmoset.

Brazilian Amazon, S of Rio Amazonas, into E Bolivia and N Paraguay. Three subspecies: *C. a. argentata*—predominantly white, pink face and ears, black tail; *C. a. leucippe*—completely pale with orange or gold tip; *C. a. melanura*—predominantly dark brown with a black face, pale hip and thigh patches. No ear-tufts.

Common marmoset

Callithrix jacchus

NE Brazil in states of Piauí, Paraiba, Ceará, Pernambuco, Alagoas and Bahia; introduced in Minas Gerais and Rio de Janeiro. Coat: elongated white ear-tufts; body mottled gray-brown; tail ringed; crown blackish with a white blaze on forehead.

Black tufted-ear marmoset

Callithrix penicillata
Black tufted-ear or Black-eared marmoset.

South C Brazil in states of Goiás, São Paulo, Minas Gerais and Bahia. Two subspecies: *C. p. penicillata*, *C. p. kuhlii*. Coat: body mottled gray; ringed tail; black face. *C. p. kuhlii* (recognized as a species by some), restricted to small area of Bahia, distinguished by brown bases to hairs on outer thighs and flank, extensive pale cheek patches and a buffy-brown crown.

Buffy tufted-ear marmoset [E]

Callithrix aurita

Remnant forests in SE Brazil in states of Rio de Janeiro, São Paulo and Minas Gerais. Coat: whitish or buffy ear-tufts; forehead ochraceous to whitish; front of crown tawny or pale buff; side of face and temples black; back agouti to dark brown or black striated as in Common marmoset; tail ringed, underparts black to ochraceous.

Geoffroy's tufted-ear marmoset

Callithrix geoffroyi

Remnant forests in SE Brazil in states of Minas Gerais and Espírito Santo. Coat: blackish-brown, with elongated black ear-tufts; forehead, cheeks and vertex of crown white; tail ringed; underparts dark brown.

Buffy-headed marmoset [E]

Callithrix flaviceps

Remnant forests in SE Brazil in states Espírito Santo and Minas Gerais. Coat: ear-tufts, crown, side of face and cheeks ochraceous; back grizzled striated agouti; underparts yellowish to orange.

Genus *Cebuella*

Pygmy marmoset

Cebuella pygmaea

Upper Amazonia in Colombia, Peru, Ecuador, N Bolivia and Brazil. Prefers floodplain forest and natural forest edge. Diet: exudate, insects, spiders, some fruits. HBL 7in; TL 7in; wt 4–7oz. The smallest living monkey. Coat: tawny agouti; long hairs of head and cheeks form a mane; tail ringed. Gestation: 136 days. Litter: 2.

Tamarins

Genus *Saguinus*

In tropical evergreen primary forests, secondary growth forest and semi-deciduous dry forests. Diet: fruit, flowers, leaf buds, insects, spiders, snails, lizards, frogs and plant exudates, such as gums, and nectar, when readily available. HBL 7–8in; TL 10–11in; wt 9–13oz. Coat: very variable. Three sections: a) hairy-faced tamarins including white-mouth tamarins (*S. nigricollis* and *S. fuscicollis*), moustached tamarins (*S. mystax*, *S. labiatus* and *S. imperator*) and Midas tamarin (*S. midas*); b) the Mottle-face tamarin (*S. inustus*); and c) the bare-face tamarins (*S. oedipus*, *S. geoffroyi*, *S. leucopus* and *S. bicolor*). Gestation: 140–170 days. Litter: 2.

Black-mantle tamarin

Saguinus nigricollis
Black-mantle or Black-and-red tamarin.

Upper Amazonia, N of Rio Amazonas, in S Colombia, Ecuador, Peru and Brazil. Coat: head, neck, mantle and forelimbs blackish-brown; hairs around mouth, sides of nostrils gray; lower back, rump, thighs and underparts olivaceous, buffy-brown or reddish. Two subspecies: *S. n. nigricollis* with black mantle, *S. n. graellsi* with agouti mantle.

Saddle-back tamarin

Saguinus fuscicollis

Upper Amazonia, W of Rio Madeira to S of Rio Amazonas and S of Rios Japurá-Caquetá and Caguán in Brazil, Bolivia, Peru and Ecuador. Coat: extremely variable amongst 14 subspecies occurring between major as well as minor river systems. All have tri-zonal back coloration which except in palest forms produces a distinct rump, saddle and mantle. The cheeks are covered in white hairs. Fourteen subspecies: *acrensis*, *avilapiresi*, *crandalli*, *cruzlimai*, *fuscicollis*, *fuscus*, *illigeri*, *lagonotus*, *leucogenys*, *melanoleucus*, *nigrifrons*, *primitivus*, *tripartitus* and *weddelli*.

Black-chested moustached tamarin

Saguinus mystax
Black-chested moustached or Moustached tamarin.

South of Rio Amazonas-Solimões between Rio Madeira and Marañon-Huallaga in Brazil, Bolivia and Peru. Three subspecies. Coat: crown and tail black (*S. m. mystax* and *S. m. pluto*) or rusty red (*S. m. pileatus*); prominent white nose and mustache; mantle blackish-brown; back, rump and outer thighs blackish striped with orange (*S. m. mystax*) or blackish-brown (*S. m. pileatus* and *S. m. pluto*); tail black; *S. m. pluto* distinguished from *S. m. mystax* by base of tail being white.

Red-chested moustached tamarin

Saguinus labiatus
Red-chested moustached or White-lipped tamarin.

Two subspecies widely separated: *S.l. labiatus* S of Rio Amazonas between Rios Purús and Madeira in Brazil and Bolivia; *S. l. thomasi* between Rios Japurá and Solimões in Brazil. Coat: crown with golden, reddish or coppery line and a black gray or silvery spot behind; mouth and cheeks covered by a thin line of white hairs; back black marbled with silvery hairs; throat and upper chest black; underparts reddish or orange.

Emperor tamarin [I]

Saguinus imperator

Amazonia in extreme SE Peru, NW Bolivia and NW Brazil. Coat: gray; elongated mustaches; crown silvery-brown; tail reddish-orange. Two subspecies *S. i. subgrisescens* has small white beard lacking in *S. i. imperator*.

Red-handed tamarin

Saguinus midas
Red-handed or Midas tamarin.

Amazonia, N of Rio Amazonas, E of Rio Negro in Brazil, Surinam, Guyana and French Guiana. Two subspecies: *S. m. midas* N of Rio Amazonas, *S. m. niger* to S in state of Pará, Brazil. Coat: black face; middle and lower back black marbled with reddish or orange hairs. Hands and feet of *S. m. midas* golden orange, those of *S. m. niger* black.

Mottle-faced tamarin

Saguinus inustus

N of Rio Amazonas between Rio Negro and Rio Japurá in Brazil extending into Colombia. Coat: uniformly black, parts of face without pigment, giving mottled appearance.

Pied bare-face tamarin [I]

Saguinus bicolor

N of Rio Amazonas between Rio Negro and Rio Parú do Oeste. Coat: head from throat to crown black and bare; tail blackish to pale brown above and reddish to orange below; large ears. Three subspecies: *S. b. bicolor* (Pied tamarin) has white forequarters, brownish-agouti hindquarters and reddish underbelly; *S. b. martinsi* has brown back and *S. b. ochraceous* is more uniformly pale brown.

Cotton-top tamarin

Saguinus oedipus

NW Colombia. Coat: crest of long, whitish hairs from forehead to nape flowing over the shoulders; back brown; underparts, arms and legs whitish to yellow; rump and inner sides of thighs reddish-orange; tail reddish-orange towards base, blackish towards tip.

Geoffroy's tamarin

Saguinus geoffroyi

NW Colombia, Panama and Costa Rica. Coat: skin of head and throat black with short white hairs; wedge-shaped mid-frontal white crest sharply defined from reddish mantle; back mixed black and buffy hairs; sides of neck, arms and upper chest whitish; underparts white to yellowish; tail black mixed with reddish hairs towards base black at tip.

Silvery-brown bare-face tamarin [V]

Saguinus leucopus

Silvery-brown bare-face tamarin or White-footed tamarin.

N Colombia between Rios Magdalena and Cauca. Coat: hairs of cheeks are long, forming upward and outward crest; forehead and crown covered with short silvery hairs; back dark brown; outer sides of shoulders and thighs whitish; chest and inner sides of arms and legs reddish-brown; tail dark brown with silvery-orange streaks on undersurface.

▲ **Unmistakable crest** of the Cotton-top tamarin of Colombia.

Genus *Leontopithecus*

Lion tamarin [E]

Leontopithecus rosalia

Lion tamarin, Golden lion tamarin, Golden-headed tamarin, Golden-rumped tamarin.

Non-overlapping and minute distributions, in primary remnant forests in Brazil. Three subspecies (considered separate species by some): Golden lion tamarin (*L. r. rosalia*) in the state of Rio de Janeiro, Golden-rumped tamarin (*L. r. chrysopygus*) in São Paulo, and Golden-headed tamarin (*L. r. chrysomelas*) in Bahia. Diet: fruits, flowers, frogs, lizards, snails, insects and plant exudates (gums and nectar) when readily available. HBL 13–16in; TL 10–15in; wt 22–25oz. The largest member of the Callitrichidae. Coat: long hairs of crown, cheeks and sides of neck form an erectile mane. Subspecies *L. r. rosalia* is entirely golden; *L. r. chrysomelas* is black with a golden mane, forearms and rump; *L. r. chrysopygus* is black with golden rump and thighs. Gestation: 128 days. Litter: 2.

Genus *Callimico*

Goeldi's monkey [R]

Callimico goeldii

Upper Amazon in Brazil, N Bolivia, Peru, Colombia and Ecuador, in dense undergrowth of non-riverine forests and bamboo forests. Diet: fruits, insects, spiders, lizards, frogs, snakes. HBL 10–12in; TL 10–12in; wt 14–24oz. Coat: black or blackish-brown, with bobbed mane, tiered pair of lateral tufts on back of crown, thick side-whiskers extending below jaws. Gestation: 150–160 days. Litter: 1.

Cooperation is Better than Conflict

Why Saddle-back and Emperor tamarins share a territory

Associations between two or more species of monkeys are common in the forests of Africa and tropical America, but why the participants join forces has not been investigated until recently. In the Amazonian region several types of mixed association can be observed, one of which involves two members of the marmoset family, the Saddle-back and Emperor tamarins. Both of these squirrel-sized monkeys live in extended family groups that include from 2–10 individuals.

In parts of southeastern Peru it is common to observe Saddle-back and Emperor tamarins together. But to find out why the two species associate it is necessary to know how contact between the two groups is maintained, and whether just one or both species actively participate in promoting the association. If each species actively joins or follows the other, a mutually beneficial interaction is implied. However, if one actively joins and follows while the other is passive, evasive or hostile, it suggests that the active joiner benefits and that its associate does not.

By following the tamarins through the forest it has been found that the associated groups maintain frequent contact through vocal exchanges and are thus able to coordinate their movements even though they are not always in sight of one another. Following periods of separation, either species may go to join the other, implying that the association is of benefit to both participants.

An unexpected finding in Peru was that the associated groups live within a common set of territorial boundaries which they defend together. Such "co-territoriality" between species had not previously been

◀ ▲ **Mutual benefit** is the basis of joint use of the same forest area by the Saddle-back tamarin (ABOVE eating fruit of *Leonia*) and the elaborately mustached Emperor tamarin LEFT, also shown FAR LEFT in its natural habitat.

recorded among South American primates (though it is well documented for species of guenon in Africa). The evidence available clearly suggested that the interaction provides benefits to both associates, but as yet there were no clues as to the nature of the benefits. The most likely areas of mutual advantage fall into three categories: the detection of predators, defense of the territory, or more efficient harvesting of food.

Both species are highly vigilant, and each responds instantly to the alarm calls of the other, so there does appear to be a roughly equal exchange of benefits in the form of anti-predator warnings. However, different species commonly respond to each other's alarms without sharing territories, so there must be some additional factor which accounts for the co-territoriality of the two species.

One possibility is that the associated species may cooperate to defend or even extend the shared territorial boundary. An opportunity to test this idea arose fortuitously when a group of Saddle-back tamarins disappeared, leaving the territory occupied only by Emperor tamarins. The family of Saddle-back tamarins that occupied the adjacent territory with a second group of Emperor tamarins then began to use both territories, switching back and forth from one to the other every few days. Although the two Emperor tamarin groups regularly confronted one another at their common border, the Saddle-back family was on friendly terms with both Emperor tamarin groups, and during clashes rested quietly on the sidelines. This observation appears to rule out the possibility of mutual reinforcement in territorial defense.

The third possibility is that the association somehow leads to more efficient exploitation of food resources. The available evidence supports this idea. Although the Saddle-back tamarin appears to have more manipulative skills and may eat more insects, both species are primarily fruit-eaters and feed on the same species of plants, often together in the same tree crown. The plant species that are most important in their diets are common small trees or vines that ripen their crops gradually over periods of weeks or months. A territory may contain 50 or more plants of a given species, virtually every one of which will be exploited by the resident tamarins. However, each plant is harvested no more than once every several days. By sharing a common territory, traveling together, and keeping in constant contact, tamarin groups can effectively regulate the intervals between visits, allowing sufficient time for the accumulation of ripe fruit, and thereby assuring that each visit will be well rewarded. If they did not share the same territory, or if they traveled independently within it, neither species would know when a given fruit tree had last been harvested, and would make frequent unrewarded trips to trees stripped of ripe fruit by the other species. Through cooperating, both species minimize the number of visits to empty trees and thus reduce their need to travel, consequently lowering their exposure to predators.

The relationship between the two tamarin species thus leads to a most unusual accommodation. In consuming the same types of fruit they are potentially the closest of competitors, yet by cooperating they attain a higher level of efficiency in harvesting the fruit crops in their territory than either species could by ignoring the other. JWT

On the Brink of Extinction

Saving the Lion tamarins of Brazil

The Jesuit Antonio Pigafetta, chronicler of Magellan's voyage around the world, referred to them as "beautiful simian-like cats similar to small lions." Though not taxonomically accurate, Pigafetta's description, based on observations made in 1519, expresses well one's first impression of a Golden lion tamarin (*Leontopithecus rosalia rosalia*)—one of the most strikingly colored of all mammals and, unfortunately, one of the most endangered.

The Golden lion tamarin has a magnificent pale to rich reddish-gold coat and a long, back-swept mane that covers the ears and frames the dark, almost bare face, and sometimes brown or black markings on the tail. There are two other subspecies (recognized as species by some authorities, see p51), both black with gold markings, and both at least as endangered as their golden relative. All three Lion tamarins are found only in remnant forest patches in the Atlantic forest region of eastern Brazil. This was the first part of Brazil to be colonized over 400 years ago, and it is now the most densely inhabited region of the country as well as its agricultural and industrial center.

The Golden lion tamarin has always been restricted to forests of low-altitude (below 1,000ft/300m) coastal portions of the state of Rio de Janeiro, with several dubious records from adjacent portions of the neighboring state of Espírito Santo. Since they are easily cleared for agriculture and pastureland, low-altitude forests are usually the first to disappear in most parts of the tropics. Habitat destruction in Rio de Janeiro increased greatly in the early 1970s, when a bridge connecting the city of Rio de Janeiro with Niterói across the bay facilitated access to remaining portions of the Golden lion tamarin's range.

Added to the major problem of habitat destruction has been live capture of tamarins for pets and zoo exhibits. Hundreds of animals were exported legally in the 1960s, to be followed by uncounted others in an illegal trade that developed after the enactment of national and international protective measures in the late 1960s and early 1970s. This trade continues to the present day, mainly to serve a local market within eastern Brazil.

On top of this, it is reported that the Golden lion tamarin is sometimes even eaten by local people in Rio de Janeiro, a sad waste of such a beautiful and internationally important little animal.

Very few Golden lion tamarins still exist in the wild. Surveys conducted over the past three years by a joint Brazilian–American

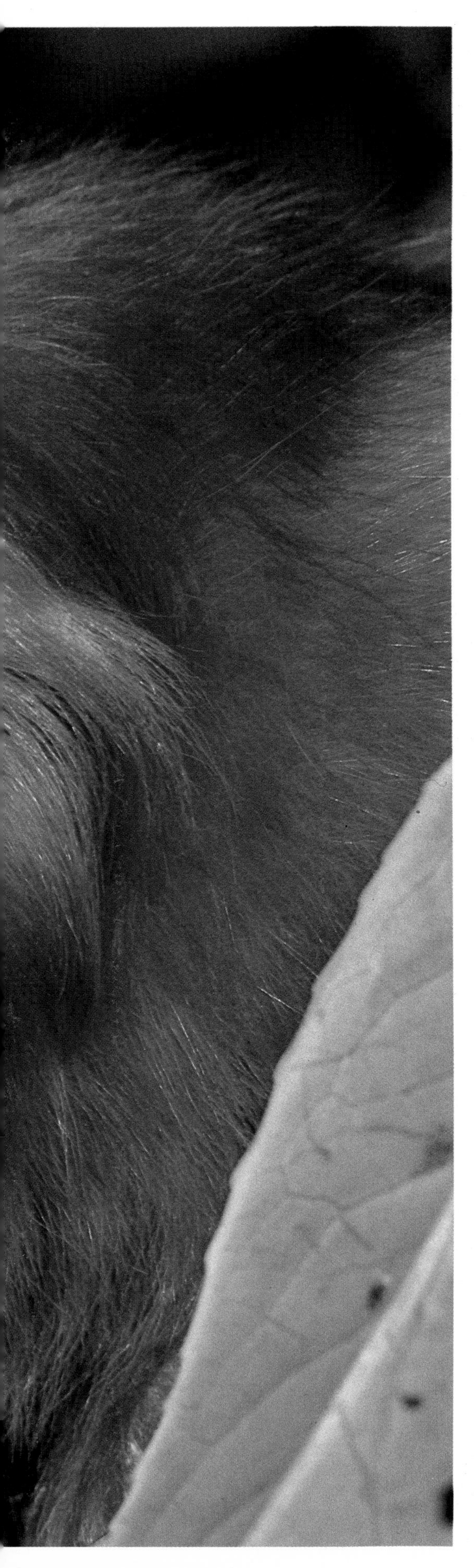

team of personnel from the Rio de Janeiro Primate Center and the World Wildlife Fund–US Primate Program have located wild populations in only two areas, one of them a stretch of forest along the coast to the south of the mouth of the Rio São João and the other the 19 sq mi (5,000 hectare) Poço d'Anta Biological Reserve established in 1974 mainly for the protection of this species. The former has already been divided into lots for beachfront housing development and appears to be doomed. The situation in Poço d'Anta is also far from satisfactory. The total population of Golden lion tamarins in the reserve is estimated to be 75–150 animals. The reserve is cut by a railroad and a road, a dam that will flood a portion of it is now being completed, poaching still takes place within its borders, and the guard force in the reserve is not sufficient to patrol it at maximum efficiency. Furthermore, only about 10 percent of the reserve is mature forest and only about 30 percent is suitable habitat for Golden lion tamarins.

A joint program involving personnel from the Rio de Janeiro Primate Center, the Brazilian Forestry Development Institute (which adminsters the reserve), the National Zoo in Washington, DC, and the World Wildlife Fund, is now being developed. Its aims are to provide detailed data on the ecology and population dynamics of the remaining Golden lion tamarins in the reserve, to restore the forest habitat, and eventually to reintroduce Golden lion tamarins into uninhabited portions of the reserve. This program must succeed if this species is to survive in the wild.

Fortunately, the species seems to be saved in captivity, thanks particularly to the effort of Dr Devra G. Kleiman of the National Zoo in Washington, DC, the Studbook Keeper for the captive colonies. Colonies in the USA and Europe now number more than 300 animals (up from 90–100 in 1972), and they continue to increase. In Brazil, the Rio de Janeiro Primate Center has about two dozen Golden lion tamarins under the supervision of Dr Adelmar F. Coimbra-Filho, Brazil's leading primatologist and the center's director. In 1980, five Golden lion tamarins from the Brazilian colonies were sent to the USA to ensure that genetic diversity is maintained, and future exchanges of animals are planned between the two countries.

▲ **Lion tamarins** are monogamous like other callitrichids, and give birth to more than one young, usually twins.

◀ **The golden mane** is common to all Lion tamarins, but the Golden lion is the only entirely gold subspecies, the others including black in their coloration.

The situation of the other two subspecies of Lion tamarin is similar. The Golden-rumped lion tamarin (*L. r. chrysopygus*) has always been restricted to the interior of the state of São Paulo, Brazil's most highly developed state, and is probably the rarest South American monkey. Much of its habitat had already been cleared by the early 1900s and none was seen between 1905 and 1970. In 1970 a remnant population was discovered in the 145 sq mi (37,000 hectare) Morro do Diabo State Forest Reserve in extreme southwestern São Paulo. A few years later, a smaller population was also found in the 5,362-acre (2,170-hectare) Caitetus Reserve in central São Paulo. These widely separate populations are almost certainly the last remaining wild representatives of this subspecies, which probably numbers no more than 100 free-living individuals.

The Golden-headed lion tamarin (*L. r. chrysomelas*) has a tiny range in the southern part of the state of Bahia, one of the few parts of eastern Brazil where reasonably large stands of forest still remain, although these are being logged and cleared for various agricultural projects. The one biological reserve there was inhabited by hundreds of squatters in 1980, and the situation continues to deteriorate.

However, there is a thriving colony of about 25 individuals of each of these subspecies in the Rio de Janeiro Primate Center. These are the only colonies of these two subspecies in existence.

Efforts are now under way within Brazil to convince the government and the general public of the importance of these uniquely Brazilian monkeys and to improve protective measures on their behalf. If these efforts fail, then all three Lion tamarins will almost certainly be extinct in the wild by the end of the decade. RAM

CAPUCHIN-LIKE MONKEYS

Family: Cebidae
Thirty species in 11 genera.
Distribution: America (Mexico) south through S America to Paraguay, N Argentina, S Brazil.

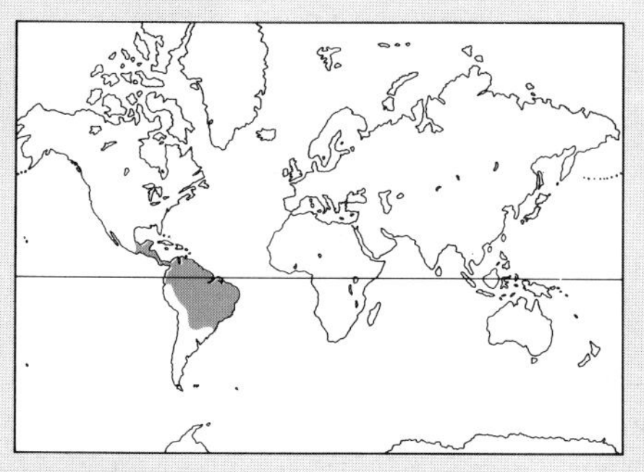

Habitat: mostly tropical and subtropical evergreen forests from sea level to 3,280ft (1,000m).

Size: from head-body length 10–14.6in (25–37cm), tail length 14.6–17.5in (37–44.5cm) and weight 1.3–2.4lb (0.6–1.1kg) in male Squirrel monkey, to head-body length 18.1–24.8in (46–63cm), tail length 25.6–29.1in (65–74cm), and weight to 26.4lb (12kg) or more in the muriqui or Woolly spider monkey. Males often larger than females but not always.

Coat: white, yellow, red to brown, black; patterning mostly around head.

Gestation: from about 120 days to 225 days, depending on genus.

Longevity: maximum in 12–25 years range for most species.

Species include:
Night monkey (*Aotus trivirgatus*).
Titi monkeys (*Callicebus*, 3 species).
Squirrel monkey (*Saimiri sciureus*).
Capuchin monkeys (*Cebus*, 4 species), including **Brown capuchin** (*C. apella*) and **White-fronted capuchin** (*C. albifrons*).
Saki monkeys (*Pithecia*, 4 species).
Bearded sakis (*Chiropotes*, 2 species).
Uakaris (*Cacajao*, 2 species).
Howler monkeys (*Alouatta*, 6 species), including **Mantled howler monkey** (*A. palliata*).
Spider monkeys (*Ateles*, 4 species).
Muriqui or **Woolly spider monkey** (*Brachyteles arachnoides*).
Woolly monkeys (*Lagothrix*, 2 species).

The monkeys of the New World have evolved into an extraordinary array of ecological, social and anatomical types, many of them unique. The 30 species of the family Cebidae of which the capuchin monkeys, *Cebus*, are the type genus include the world's only nocturnal monkey, some of the world's brainiest non-human primates, and the only primates with prehensile tails. They feed on everything from insects and fruits to leaves, seeds, and even other mammals. One species occurs only in a small and isolated mountain range (Yellow-tailed woolly monkey), while others have spread throughout tropical South America (eg Night monkey and Brown capuchin). Their social organizations range from strict monogamy to large polygamous groups.

In spite of their broad range of adaptations, cebid monkeys share some common features. The family is set apart from other primates by the wide form of the nose (specifically, of the septum that separates the nostrils), its absence of cheek pouches, and its tooth formula (I2/2, C1/1, P3/3, M3/3 = 36). They are mostly found in tropical and subtropical evergreen forest, although some have adapted to elevations as high as 9,900ft (3,000m) and to forests with marked dry seasons. They live almost exclusively in trees, but some species will descend to the ground to play (White-fronted capuchin), look for food (Squirrel monkey), or travel between patches of woodland. Nevertheless, none of the cebids show obvious specializations for life on the ground, as do so many of the Old World primates. All the cebid monkeys that have

► **Starved-looking** even when in the best of health, the Red or White uakari has almost hairless crown and facial skin whose coloration fades when kept out of sunlight. Uakaris have a short tail.

▲ **Prehensile tail** of the Black-handed spider monkey is used as a fifth limb and can support the whole weight of the animal. Spider monkeys have long arms and flexible shoulder joints for swinging along branches (brachiating).

▼ **Male** (RIGHT) **and female Black howler monkeys** have different colored coats.

been tested have some color vision, usually fairly acute, although they have poor sensitivity to the red end of the color spectrum. Fur colors tend to be white, yellow, red to brown, and black, and most of the color patterning occurs around the head.

Species such as the Squirrel monkey, which move by leaping, have thighs that are shorter relative to the lower leg, thus allowing more force to be developed in the jump, than species that more commonly clamber, such as the howler monkeys. Others, like spider monkeys, that swing hand over hand below branches have relatively long arms and extra mobility in the shoulder joint. The very flexible, prehensile tail of the larger cebids is used like a fifth limb. They can use it to grab on to branches for safety and can even hang from it to feed near the tips of branches.

Each species has a diet that matches, and may be limited by, its anatomy, in particular the forms of the jaws, teeth and gut. Leaf-eating Mantled howler monkeys have relatively broad flat teeth, very deep lower jaw bones, and a relatively large gut occupying one-third of their body volume, because they must process large amounts of tough plant material (see p72). The insect-eating Squirrel monkey has sharp, narrow teeth to macerate insects quickly, and a short, simple gut occupying less than one-sixth of its body volume, which is all that is needed to absorb this nutritious food. Even within one genus, the Brown capuchin differs from the White-fronted capuchin in having a deeper lower jaw and more massive jaw muscles. The Brown capuchin eats larger fruits and more vegetation than the White-fronted species. The bearded saki monkeys derive most of their diet from the hard seeds inside fruits, which they position in the distinctive gap between the canine and premolar teeth before cracking them open.

These characteristic adaptations of form and ecology are closely related to body size. The smallest and lightest species can most easily leap from branch to branch without risk of injury, which may explain their lack of a prehensile tail. The small species also inevitably have small, weak jaws, which limit the size and hardness of the fruits which they can efficiently eat. Small size also means a low demand for energy and protein, so that small species can afford to live on foods that are relatively scarce but easily processed, such as insects and ripe fruits. Finally, the small gut and short digestive times of the smaller cebid monkeys are adequate for easily digested foods, but not adapted for tough plant materials like mature leaves. Large monkeys of course have the opposite advantages and disadvantages: high risk of injury from falling, strong jaws, high energy demand, and long digestive times.

Although cebid monkeys coexist with many non-primate competitors such as birds and bats, they compete directly with other cebid monkeys. It is common for as many as five species to feed on one tree species or even in the same tree. When several species feed in a single tree, physical clashes determine which has precedence. Usually the smaller or less agile species are evicted. Their vulnerability to being displaced from fruit trees may have contributed to the specialized life-styles shown by the smaller species. Titi monkeys are able to eat green fruit before it is palatable to the larger monkeys. Night monkeys feed at night in the same trees that are dominated during the day by big species. Squirrel monkeys find safety in numbers that are too great for larger monkeys, living in smaller groups, to chase out of a tree.

Even when different species do not share the same fruit trees, it may be as a result of effective competition between them. The species which exploits a given resource most efficiently will generally force other species to choose different foods. Thus the greater ability of large monkeys to eat big fruits and to digest tough leaves may have accelerated the tendency of smaller monkeys to use other resources. These pressures may explain why bearded sakis are specialized for opening the seeds inside fruits, and why uakaris, which eat many of the same foods as other medium-sized monkeys, are found only in a restricted habitat, flooded swamp forest, where other cebids are absent.

Different species within one genus are likely to compete severely because they are more likely to be similar in anatomy and ecology. That competition for food has influenced the evolution of cebid monkeys is strongly suggested by the fact that in any one place it is rare to find more than one species of the same genus. In the few examples of such overlap, such as between the Brown capuchin and other capuchin species, there is some evidence that they have reduced competition by eating less similar foods where they coexist than where they occur alone.

Each species of cebid monkey also has its distinctive social structure. Most smaller cebids of 1.5–3.3lb (0.7–1.5kg), including the Night monkey, titi monkeys, and probably also sakis, are monogamous. The pair bond is maintained both by friendly interactions between the adult couple and by active aggression toward other members of the same species and sex. In monogamous species, juveniles may stay with their parents for up to a year or two after the birth of the next infant, so that groups of up to five animals may be seen. An exception among the smaller species is the Squirrel monkey, which lives in large groups, typically containing 30–40 animals, with up to a dozen reproductive females and several adult males. Females tolerate females, but are not very tolerant of males and often gang up on individual males within their group to drive them away from their vicinity. During the breeding season, males may have spectacular fights that leave the loser injured or weakened. A female often mates with several males during her brief period of receptivity.

All the larger cebids (4.4–20lb/2–9kg) typically live in groups of at least five

5

▲ **Representative species** of capuchin-like monkeys, showing movement among the trees. In most species males are slightly larger than females (shown here); in a few species there are color differences between the sexes.
(1) White-faced or Guianan saki (*Pithecia pithecia*) (male has matt of buff hair on top of head). (2) Red uakari (*Cacajao rubicundus*); the two uakaris are the only short-tailed primates (other than man) in the New World. Both (3) the Dusky titi (*Callicebus moloch*) and (4) the Squirrel monkey (*Saimiri sciureus*) move chiefly by leaping and have tails that are not prehensile. (5) Brown capuchin (*Cebus apella*). (6) The Night monkey (*Aotus trivirgatus*). (7) The female Black howler monkey (*Alouatta caraya*) is olive-buff. As in other larger cebids, the tail is prehensile, naked beneath the tip for better gripping. In (8) the Black-handed spider monkey (*Ateles geoffroyi*), the tail may be used for picking up small objects (eg food items). (9) Humboldt's or Smokey woolly monkey (*Lagothrix lagotricha*).

animals and are more or less polygamous. At low population densities, howlers and capuchins often live in harems in which one male monopolizes 1–3 females. At higher population densities, these same species occur in larger groups of 7–20 animals that include several adult and subadult males. Such large, multi-male groups are normal for bearded sakis, uakaris, woolly monkeys and muriqui or Woolly spider monkey. In these species, it seems that females usually mate with the dominant male or males, but solicit and mate with subordinate or subadult males as well. Spider monkeys (and to a lesser extent woolly monkeys and the Woolly spider monkey) have variable social groupings in which individuals daily join and leave subgroups of changing size and composition. Only for a few weeks a year do all 20 or so members of a spider monkey group gather together in one place, but all group members continue to recognize and tolerate one another over long periods.

Group size seems to depend to a large extent on the productivity and abundance of the foods typically eaten by a given species. Most species that live in small groups feed on small, scarce, and scattered resources, such as insects, small fruit crops in vines, and newly emerged leaves of vines and bamboo. Species with large groups typically use large, abundant, and clumped or isolated resources, such as fruits on large fig trees. This relation holds true even within one species of, for example, spider monkey, forming subgroups that vary in size: the smallest subgroups of 1–3 animals occur when food trees are small and scattered, and the largest of 7–20 animals when food trees are large and clumped. The exception to this general pattern is the Night monkey, with small monogamous groups, but feeding in large, clumped food trees.

How a species uses its home range also depends to a large extent on how its food resources are distributed. Titi monkeys, for example, feed in small, scattered food trees and use their home range in a very even fashion, visiting each part of it every few days. Squirrel monkeys and White-fronted capuchins depend on large, clumped or rare trees and their large groups travel their home range in an uneven way, spending most of their time in a small sector until the fruit there is exhausted, then moving on to find new sources.

These patterns of home range use in turn affect how neighboring groups of a species interact. Those cebid monkey species that exploit their home range in a very even way are usually territorial, and may defend their range with loud dawn calls, as given by titi and howler monkeys (see pp370, 372). At the opposite extreme, different groups of Squirrel monkeys seem to overlap completely and usually show no overt aggression toward each other even when feeding in the same fruit tree. Other species show intermediate patterns—Brown capuchins will fight other groups of its species only if a fruit tree is at stake, while White-fronted capuchin groups tend to avoid each other whenever possible.

The relation between large group size and abundant or large food sources may be caused by competition for food between members of a social group. Within their groups of 3–15 or more, Brown capuchins often fight over food (see box). Individuals which cannot win aggressive confrontations, or are not tolerated by other group members, suffer markedly reduced feeding success. Fighting is most common in small fruit trees, or when food is scarce. At these times some group members may forage alone or in smaller groups.

However, there are also compensating benefits to large group size. For instance, individuals in large groups probably suffer less predation. The more eyes and ears there are to watch for predators, the lower the chance that a predator attack will be successful. Throughout Central and South America there are a variety of large hawks and eagles, some of them specialized to eat monkeys. In southeastern Peru, each capuchin group is attacked by an eagle once every two weeks on average, and there are less serious threats up to several times a day. Brown capuchins are so wary that they consistently take alarm at completely harmless birds that fly by.

In the absence of predation, social groups could be relatively small. The lack of major nocturnal predators may allow the Night monkey to live in small groups, even though they feed in large, clumped food trees (see p68). Spider monkeys may be too big for eagles to kill, and this immunity may facilitate the formation of their characteristic small foraging subgroups.

Other advantages to group life include communal finding and defending of food trees. Although individuals of many species know in advance what trees will be in fruit and where they are, some portion of their diet comes from unpredictable or small sources which they find by chance. In either case, extra individuals in the group can increase the pool of food available to the group as a whole by sharing knowledge about fruit trees. In fact, at least one species, the Brown capuchin, has highly distinctive loud whistles which are given when a group member finds or begins to use a rich source of food. Larger capuchin groups also usually win fights for possession of fruit trees.

There are other, more specific advantages to members of large groups. Squirrel monkey females receive no help from males in rearing the young, but they do have other female "friends" that help carry and watch over an infant while its mother forages. The "friends" that do not have their own young are the ones most able to help a mother, and often are her own offspring from previous years. A male howler monkey deciding whether to join a new group may prefer one which has a relatively large number of reproductive females. For a juvenile White-fronted capuchin, having a large number of young group mates to play with may be important to its social and physical development.

Several cebid monkeys actively form mixed groups with members of other species. A great variety of stable associations has been recorded, most notably between Squirrel monkeys and capuchins, Squirrel monkeys and uakaris, even capuchins and spider monkeys. Observations in Peru and Panama suggest that the benefits are often one-sided in such mixed associations. For instance, it is almost always the Squirrel monkeys that join up with the capuchins, and capuchins usually determine the direction of movement of the combined group.

Social Station in Brown Capuchins

As a group of Brown capuchins spreads out in search of food, it is possible to predict which position each monkey will take up as it forages. Each individual's position depends upon its foraging success when food is scarce. If there is squabbling at a food tree it is the most dominant male and those whom he tolerates that almost always feed first and for longest. The subordinate individuals whose proximity the dominant male will not tolerate must wait until the dominant's entourage has left the tree.

The desirability of a given position in a foraging group can be rated by the extent to which it offers access to food and safety from predators. Individuals securing a central position benefit from the watchfulness of their neighbors whose presence also shelters them from a direct attack. However, success in foraging may be highest at the leading edge of a moving group, where new resources are being discovered. But if an individual can parasitize the efforts of 8–12 members who scout, then the best position is behind the leading edge, but close enough to monitor the success of individuals at the front.

Group members keep a wary eye on the dominant male and adjust their own position according to his tolerance of them. The dominant male occupies the central area just behind the leading edge, where both his feeding success and protection from predators are greatest. He is very tolerant of infants and young juveniles, and these tend to follow him, just behind the center of the group. Females tolerated by the dominant male, plus their older offspring, often form the leading edge of the group, trading off foraging success for exposure to predation. Finally, those females and males not tolerated by the dominant male are found on the periphery of the foraging group, at the trailing edge, or even further away from the rest of the group. Their priority seems to be to avoid aggression, even at the expense of increased time spent looking for food and/or greater risk of attack by predators.

CHJ

◀ **Pair of Guianan or White-faced sakis.** Like many other New World monkeys, these live in monogamous pairs or small family groups. Only the male has the white face.

▼ **The muriqui** or Woolly spider monkey is the largest and most ape-like of the New World monkeys. It also appears to be the most endangered neo-tropical primate species. Found only in the Atlantic forest region of southeastern Brazil (though once abundant in this area) its numbers have fallen to just 200–250 muriquis in seven small remnant forest areas. Three of these forests are government reserves and four are privately protected, but none can be considered entirely secure against further forest destruction over the long term.

Since the muriqui is the largest mammal entirely restricted to Brazil and also a single-species genus with no other close relatives, it is a most appropriate symbol of the Brazilian conservation movement.

Adult muriquis weigh at least 26lb (12kg) and perhaps as much as 33lb (15kg), and can measure almost 5ft (1.5m) from head to tail tip.

Because capuchins have significantly smaller home ranges than Squirrel monkeys—0.4–0.8sq mi (1–2sq km) compared with over 1.5sq mi (4sq km), they probably know better the wheareabout of fruiting trees. The Squirrel monkeys arrive more quickly at local food trees by following the capuchins than if they forage alone. This explains why such associations form only between ecologically similar species, and also why only one of the species actively maintains the mixed group. (For two species of a different family, the marmosets and tamarins, that derive mutual benefit from sharing a territory, see p52.)

Regardless of their social organization, most species have a similar diversity and range of calls. This basic set includes some forms of trills, squeaks, grunts, hoots, barks, and whistles. Nevertheless, distinctive differences arise because each species uses only a certain number of the possible basic call types, which it combines in characteristic ways, and employs in different contexts. A striking example is the different forms of the call used for contacting and intimidating other groups: Night monkeys use low resonant hoots, howlers have loud roars, Brown capuchins give whistles, and titis perform a complex sequence of gobbles and pumping notes.

Some social displays are common to many, if not all, cebid genera. Most species show some form of threat behavior to other members of the same species, and even toward other species and predators. The usual posture is with the mouth opened and teeth bared to expose the canines. Not surprisingly, this display seems to be absent in titi monkeys, which have small canines, and Night monkeys, which fight at night. Along with the open mouth display, a monkey often shakes a nearby branch, bounces on it, or tries to break it off, a hazard to the biologist lurking below.

Grooming is the most common friendly behavior shown by cebids. Often an individual approaches another and lies down in a characteristic posture on the branch next to it, or even right on top of its feet. Occasionally an animal will take the initiative in grooming another. The groomer usually grooms areas difficult for the other monkey to reach or see, such as the back of the neck, crown, and hind limbs. An individual grooms for only a few minutes at a time, then turns around and solicits grooming from its partner. According to the old rule "I'll scratch your back if you scratch mine," the pair usually exchanges grooming bouts several times; the sequence often stops when one of the partners refuses to return grooming.

The behavior of infant monkeys differs from adults' in many ways. They are clearly less coordinated and less aware of the appropriate occasion on which to behave in a particular way. Some juvenile behavior is almost entirely absent from the adult repetoire, and vice versa. It seems that many adult behavior patterns have to be learned, or at least practiced, before an individual can perform them correctly. Since their food supply is assured, infants spend much of their time playing, either exploring their environment or chasing, grappling, wrestling, and jumping on other juveniles. Both within and across species, the more playmates a juvenile has, the more likely it is to play. Monogamous species with small groups tend to show relatively little play for the amount of time they spend resting, whereas in Squirrel monkeys, the infants often form a completely distinct subgroup which plays continuously whenever the group moves slowly (see p66).

A great deal still remains to be learned about cebid monkeys. A number of species in several genera are known only from museum specimens. Opinions on the division of the family into subfamilies, for example, are constantly changing. Only a few species have been studied for over five years, yet long-term studies are essential to understanding the behavioral flexibility and development of individuals and the relation between environment and behavior.

Over one-third of the species of cebid monkeys are listed as endangered or vulnerable to extinction. The Woolly spider monkey (or muriqui) is the most endangered (see LEFT). In the Amazon basin, spider monkeys and woolly monkeys have been shot for food out of large portions of their former ranges. These species are especially vulnerable because of their slow maturation, low reproductive rate, and dependence on mature rain forest that is being cut down for farms and wood pulp at the rate of 35,000 acres (14,000 hectares) a day. Smaller cebids, like the titis, Night monkey and Squirrel monkey, can adapt more easily to habitat loss and change. But even for these the impending destruction of mature rain forest throughout South America will prevent scientists from ever observing them in a truly natural state. Species that are restricted to tiny ranges, like the Yellow-tailed woolly monkey, or are naturally fairly scarce, like all the species of uakaris, may become extinct before mankind can learn of their unique adaptations.

CHJ

THE 30 SPECIES OF CAPUCHIN-LIKE MONKEYS

Abbreviations: HBL = head-and-body length; TL = tail length; wt = weight.
[E] Endangered. [V] Vulnerable. [I] Threatened, but status indeterminate.

Classification of cebids

The constant revision of the taxomomy of the cebids is a reflection of how little is known about these New World monkeys. Many species have not been studied since they were first discovered, and their distributions are known only from a few collecting localities. The scientific criterion for distinguishing a species—whether it can interbreed extensively with other animals—cannot be applied to many cebids because they do not overlap in range. In other instances, there are important differences in the genetic material between forms that cannot reliably be distinguished by color. There is even debate about whether all the cebids, or just *Saimiri* and *Cebus*, should be united with the marmosets and tamarins in a single family. The treatment here is the more usual one of keeping them separate. Where the species status of a distinct form is debated, this is noted in the introductory remarks to the genus.

Genus *Aotus*

Night monkey
Aotus trivirgatus
Night monkey, Owl monkey, douroucouli.

From 81°W in Panama south to N Argentina, except Guianas, Uruguay, Chile, E Brazil. Savanna and dry to rain forests, from sea level to over 9,800ft. Primarily eats fruits, leaves, and insects; vertebrates and eggs eaten occasionally. Chiefly nocturnal. Monogamous, in groups of 2–5. Sexes similar in size. HBL 9–19in; TL 9–16in; wt 2–3lb. Coat: grizzled brown, gray, or reddish on back, back of head, and limbs; underside buff-white to bright orange; head distinctive, with triagonal white patches above large eyes, and three black stripes, between and either side of eyes, converging on crown. Populations with similar coats may vary in chromosome number and pattern. (Other authorities recognize 2, 3, even 9 distinct species.) Gestation: 133 days. Litter size: 1, rarely 2. Birth frequency: 1 per year, births seasonal in some areas.

Titi monkeys

Genus *Callicebus*

In disturbed to mature, moist to rain forest; from sea level to about 3,300ft. Prefer understorey up to 33ft. Feed on fruits, often unripe, insects and leaves. Coat dense and long. Canines short relative to other cebids. Males may be slightly bigger than females. Diurnal. Monogamous, in groups of 2–5. Gestation: not certain, probably over 4 months. Litter size: always 1. Births once a year, usually in early rainy season.

Dusky titi
Callicebus moloch
Dusky or Red titi.

Colombia and W Venezuela south to S Bolivia, E of the Andes to Rio Negro N of the Amazon and to Maranhão (Brazil) S of the Amazon. Prefers swampy, flooded, disturbed habitats. More leaf-eating than Yellow-handed titi, and can live on leaves and green fruits alone for short periods. HBL 10–16in; TL 12–20in; wt 1.5lb. Coat: gray to red-brown; belly same as back, or sharply defined orange-red; forehead may contrast, with brighter red, gray, or black band. Tip of tail pale cream or buff.

Yellow-handed titi
Callicebus torquatus
Yellow- or White-handed titi, Widow monkey, Collared titi.

N of Amazon and Rio Marañón (Peru) to Orinoco and the Guianas; S of Amazon to Rio Purus in Brazil. Prefers unflooded, mineral-poor, sandy-soil forests. Eats more insects than Dusky titi, but over 75% of diet is fruit. HBL 12–18in; TL 15–20in; wt about 2lb. Coat: red-brown to black; forehead face, throat, and tip of tail may have contrasting white-cream to orange fur; fur on hands pale cream; forearms, legs, and feet black; tail dark brown to black.

CONTINUED ▶

▶ **Humboldt's woolly monkey** (or Smokey woolly monkey) young.

Masked titi V
Callicebus personatus

E coast of Brazil from Bahia S to São Paulo (no overlap with other species). Diet not known. HBL 12–16in; TL 16–22in; wt ? Coat: dark brown with buffy tips on hairs; belly may be more red-brown; facial fur dark brown or black; hands and feet always black; tail reddish-brown.

Genus *Saimiri*

Squirrel monkey
Saimiri sciureus

C America from Costa Rica to Panama (80°W), where sometimes called a distinct species, *S. oerstedii* E; in moist forests throughout N and C South America to Bolivia and Brazil (except Atlantic forests). Disturbed to mature upland, river-edge, and mangrove forest. From sea level to about 3,300ft elevation. Mostly eats fruit and insects, but may survive on insects. Diurnal. Polygamous, in groups usually of 30–40 individuals. Males larger than females. HBL 9–12in (female), 10–14in (male); TL 14–17in (female), 14–18in (male); wt 1–2lb (female), 1–2lb (male). Fur short, generally yellow to yellow-green on back and limbs. Underside paler yellow to white. Face, throat, and ears white. Muzzle and tip of tail black. Crown and sideburns from darker green to black. Gestation: 152–168 days. Litter size: 1. Birth frequency: 1 per year, very seasonal and synchronous within a group.

Capuchin monkeys

Genus *Cebus*

In many habitats from dry to rain forest, from sea level to over 8,200ft, also riverine forests, swamps, mangroves, seashores. Prefer understorey to midcanopy, but often come to ground to forage or play. Mostly eat ripe fruits and insects, but may use unripe fruits, vegetation, flowers, seeds, roots, other invertebrates (snails, spiders), young and adult vertebrates (especially nestling birds), cultivated crops. Fur of medium thickness and length. Males bigger in body and relative canine size. Diurnal. Polygamous, in groups of 3–30 with 1 to several males. Adult male/female sex ratio in groups is about 1:1 in small groups, up to 1:3 in large groups. Gestation: about 150 days. Litter size: 1 (twins very rare). Births usually once every 2 years, but may be faster if previous infant dies. Births often seasonal, in early rainy season, but not as well synchronized as the Squirrel monkey. Two species groups exist, one including only *C. apella*, which overlaps with members of the other group; the other 3 species do not overlap or barely meet.

Brown capuchin
Cebus apella
Brown, Tufted, or Black-capped capuchin.

Throughout South America E of the Andes in subtropical and tropical forest, except Uruguay and Chile. Prefers moister forests than other capuchins, ranges to higher elevations (8,800ft); eats more vegetation and larger fruits; groups generally smaller, 3–15 animals. HBL 13–19in (female), 12–22in (male); TL 15–19in (female), 15–22in (male); wt 5–7lb (female), 8–8.5lb (male). Coat: coarser than other species, generally light to dark brown, paler on belly; extremities of limbs always black; facial pattern varies, but usually pale skin shows through sparse whitish fur on face, surrounded by distinctive black sideburns extending to chin; black cap forming a downward pointing triangle on forehead, may extend to form "tufts" each side of crown.

White-faced capuchin
Cebus capucinus
White-faced or White-throated capuchin.

C America from Belize south to N and W Colombia. In dry to wet forests up to 6,900ft, and in mangroves. Group size 10–24. HBL 12–16in (female), 13–18in (male); TL 16–18in (female), 16–20in (male); wt 4–6lb (female), 5–8lb (male). Coat: fur on upper limbs, sides of body and belly pale cream to white, grading to black on back and extremities of limbs; distinct cap nearly black, rest of facial fur white; longer forehead and crown hairs on older individuals may form a "ruff."

Weeper capuchin
Cebus nigrivittatus
Weeper or Wedge-capped capuchin.

S America N of Amazon and N and E of Rio Negro in Brazil through the Guianas to central Venezuela. Habitat similar to *C. capucinus*. Group size 10–33. HBL 15in (female), 15–18in (male); TL 18in (female), 17–19in (male); wt 5lb (female), 6lb (male). Coat: similar in pattern to White-faced capuchin, but less contrast between light and dark colors on body, rarely becoming black; crown patch narrow, coming to distinct point on forehead.

White-fronted capuchin
Cebus albifrons

W and S of range of Wedge-capped Capuchin, from Venezuela and Brazil W of Rio Negro and Tapajós throughout upper Amazon to Bolivia; also W of Andes in Ecuador, and in N Colombia and W Venezuela; Trinidad. May prefer less disturbed, moister forests than other capuchins. Group size 7–30. HBL 13–16in (female), 14–17in (male); TL 16–20in (female), 16–19in (male); wt ? Coat: very similar to Wedge-capped capuchin, but never dark brown to black; cap paler and broad, covering most of top of head; limb extremities and tail tip usually paler, unlike other *Cebus*.

Saki monkeys

Genus *Pithecia*

Prefer mature high-ground forest, but also in disturbed forests and savannas; rarely swamp forests. From sea level, upper limit unknown. Prefer lower and middle canopy levels. Feed on fruits, including immature seeds, and some leaves. Coat coarse and long, especially on tail. Gap between canines and premolars. Males slightly larger than females. Diurnal. Probably monogamous or in extended family groups of 2–5 individuals. Gestation: 163–176 days (known only for Guianan saki). Litter size 1. Birth interval and seasonality not known.

White-faced saki
Pithecia pithecia
White-faced or Guianan saki.

N of Amazon, E of Rios Negro and Orinoco, in lowland and montane forests. HBL 13in (female), 13–15in (male); TL 13–17in (female), 14–1·7in (male); wt 3–4lb (female), 4–5lb (male). Coat: in male black except for white to reddish forehead, face, and throat; in female brown to brown-gray above, paler below, with white to pale red-brown stripes from eyes to corners of mouth.

Red-bearded saki
Pithecia monachus
Red-bearded or Monk saki.

Upper Amazonia from Rios Juruá and Japurá-Caquetá (Brazil) W to Andes, S to about 8°S, in lowland forests. HBL 16in (female), 16–17in (male); TL 18in (female), 18–19in (male); wt 5lb (female), 5.5lb (male). Coat: grizzled brown-gray hands and feet paler; beard and underside reddish; males have mat of buffy hair on forehead and crown.

Black-bearded saki
Pithecia hirsuta

Distribution as Red-bearded saki but farther E, to Rio Tapajós, and S to N Bolivia (about 13°S). HBL 15–18in (female), 15–17in (male); TL 16–19in (female), 17–20in (male); wt? Coat: grizzled brown-gray, with paler hands and feet, and blackish beard and underparts.

Buffy saki
Pithecia albicans

Between lower Rios Juruá and Purus in Brazil, restricted to forests. HBL 14–16in (female), 16in (male); TL 16–18in (female), 16–17in (male); wt ? Black back and tail. Limbs, ventral fur, and head ruff buff to red.

Bearded sakis

Genus *Chiropotes*

From near sea level to mountain forests. Prefer upper canopy, rarely in understorey. Feed primarily on seeds, often unripe, and fruits; leaves eaten rarely. Coat of medium length, dense except for bushy tail. Canines and premolars separated by a space. Sexes similar in size. Diurnal. Probably polygamous; in groups of 8–30 individuals, with more than one male. Gestation: about 5 months. Litter size: 1. Birth frequency unknown, but births seasonal in some areas at start of rainy season.

Bearded saki
Chiropotes satanas
Bearded or Black saki.

N of Amazon, E of the Rio Negro and S of Rio Orinoco into the Guianas, and S of Amazon from Rio Tocantins E to Atlantic. Exclusively in high ground, mature forests. HBL 18in (female), 16in (male); TL 14in (female), 14–15in (male); wt 6lb (female), 6lb (male). Coat: mostly black, with back and shoulders light yellow-brown to dark brown; distinctive swollen temples covered with fur, and beards roughly as long as the face.

White-nosed saki V
Chiropotes albinasus

S of Amazon between Rios Xingu and Madeira, to near Bolivia. In both high ground and permanent swamp forests. HBL 15in; TL 16in; wt 5lb (female), 7lb (male). Coat: entirely black except for white nose; temporal swellings and beard less developed than Bearded saki.

Uakaris

Genus *Cacajao*

Apparently prefer lowland swamp forest, if not limited to it. Eat mostly fruit and leaves, some seeds and insects. Coat long and coarse. Tail very short, only about one-third body length. Canines and premolars separated by a space. Males slightly larger than females. Diurnal. Probably polygamous in groups of 15–30 or more, with several adult males. Gestation: about 6 months. Litter size: 1. Birth frequency unknown in wild, births seasonal in captivity (semi free-ranging conditions).

Red uakari

Cacajao rubicundus
Red or White uakari.

N of Amazon from Rio Japurá (Brazil) W to Andean foothills; S of Amazon from Rio Juruá (Brazil) to Río Huallaga (Peru). HBL 14–17in (female), 17–19in (male); TL 5–6in (female), 6–7in (male); wt 8lb (female), 9lb (male). Coat: entirely white to chestnut-red; forehead and crown sparsely haired or bald; facial skin pink to scarlet.

Black uakari [V]

Cacajao melanocephalus
Black or Black-headed uakari.

N of Amazon, E of Rio Japurá to Rios Negro and Branco. Dimensions unknown. Coat: black; hindlimbs and tail partly chestnut-brown, sometimes with yellow in tail; facial skin black, fur extends over crown to forehead.

Howler monkeys

Genus *Alouatta*

In undisturbed dry to rain forests, wooded savannas. From sea level to about 8,200ft, but usually below 3,300ft. Prefer lower to middle canopy level, but descend to understorey to feed, and travel on ground when necessary. Eat mostly fruit, but with high proportion of unripe fruit. Diet 40 percent mature or young leaves. Fur medium length, sparse on belly. Male larger than female. Diurnal. Polygamous, in groups of usually 4–22, with 1 or more males. Gestation: 180–194 days. Litter size: 1 (twins rare). Birth interval: 1–2 years. Births not markedly seasonal, or with two birth peaks per year.

Mexican black howler [I]

Alouatta villosa
Mexican black or Guatemalan howler.

Yucatan Peninsula (Mexico), N Guatemala, Belize. HBL 20–21in (female), 24–25in (male); TL 25–26in (female), 26–28in (male); somewhat heavier than Mantled howler. Coat: black, base of hairs reddish-brown.

Mantled howler

Alouatta palliata

From S Veracruz, Mexico, S to N tip of Colombia on both sides of Andes, continuing on W side to Ecuador. HBL 14–24in (female), 18–25in (male); TL 22–26in (female), 20–26in (male); wt 14lb (female), 17lb (male). Coat: black or brown, mixed on back with golden brown; flank fringe yellow-brown.

Red howler

Alouatta seniculus

From N Colombia and Venezuela S to Amazon, including the Guianas: S of Amazon from Andes east to Rio Madeira, S to C Bolivia. A male that takes over a new group may kill young infants present. HBL 18–22in (female), 19–28in (male); TL 20–28in (female), 19–29in (male); wt 10lb (female), 14lb (male). Coat: orange-brown.

Red-handed howler

Alouatta belzebul

S of Amazon from Rio Madeira E to Maranhão and Ceará. HBL 16–25in (female), 22–25in (male); TL 18–27in (female), 22–27in (male); wt 12lb (female), 16lb (male). Coat: black or blackish-brown with yellow to reddish hands, feet and tip of tail.

Brown howler [I]

Alouatta fusca

E coast of Brazil from Bahia S to Rio Grande do Sul. HBL 17–21in (female), 20–25in (male); TL 19–22in (female), 19–26in (male); wt? Coat: brown.

Black howler

Alouatta caraya

From headwaters of Rios Madeira and Tapajós, S to S Brazil, Paraguay, E Bolivia and N Argentina. HBL 20in (female), 24–25in (male); TL 21–24in (female), 24–26in (male); wt 14lb (male). Coat: all black in male, olive-buff in female.

Spider Monkeys

Genus *Ateles*

Mature, tall, moist to rain forest. From sea level to 7,500ft. Prefer middle to upper canopy, often moving into emergent trees. Move, suspended by hands and tail, hand over hand below branches, also quadrupedally. Eat fruit and leaves but prefer more ripe fruit than howler monkeys. Fur short, not dense. Diurnal. Polygamous, in groups of about 20, but generally only 2–8 individuals move together at one time. Gestation: 225 days. Litter size: 1. Birth frequency: 1 every 2–3 years. Births not notably seasonal.

Black spider monkey [V]

Ateles paniscus

N of Amazon, E of Rios Negro and Branco through the Guianas. S of Amazon, in upper Amazonia from Rio Ucayalí E to Rio Juruá and S to headwaters of Río Madeira (Bolivia). HBL 16–22in (female), 15–19in (male); TL 29–36in (female), 25–32in (male); wt 19lb (female), 21lb (male). Coat: all glossy black.

Long-haired spider monkey [V]

Ateles belzebuth

From N Colombia S; from upper Orinoco drainage and headwaters of Rio Negro, in upper Amazonia to N Peru; isolated populations S of the Amazon, E of Rio Tapajós (Brazil) to Amazon delta, also N Venezuela. HBL 13–23in (female), 16–20in (male); TL 24–35in (female), 27–32in (male); wt 13lb (female). Coat: black, or dark brown with paler brown to white underparts, hindlimbs, and base of tail; white to yellow-brown triangular patch on forehead.

Brown-headed spider monkey [I]

Ateles fusciceps

From S Panama through N Colombia, on W side of Andes to Ecuador. HBL 18–22in (female), 14–23in (male); TL 24–32in (female), 25–28in (male); wt ? Coat: all black with white mustache and chin whiskers, or brownish-black with olive to yellow-brown crown cap.

Black-handed spider monkey [V]

Ateles geoffroyi

Throughout C America from Veracruz (Mexico) to W Panama. HBL 13–20in (female), 15–19in (male); TL 27–33in (female), 23–32in (male); wt 17lb (female), 16lb (male). Coat: variable—golden brown, red to dark brown; hands and feet black.

Genus *Brachyteles*

Muriqui [E]

Brachyteles arachnoides
Muriqui or Woolly spider monkey.

In moist rain forest of SE Brazil from Bahia to São Paulo. From sea level to 3,300ft. Has been seen to eat fruit, leaves and flowers. Fur thick and dense, similar to woolly monkeys (*Lagothrix*). Lives in multi-male troops of 25 or more. Diurnal. Single infant born. In danger of extinction. HBL 18–22in (female), 18–25in (male); TL 29–31in (female), 25–29in (male); wt to 26lb or more. Coat: uniform pale gray to brown, sometimes with extensive yellow in males.

Woolly monkeys

Genus *Lagothrix*

Usually in mature moist to rain forest from sea level to over 8,200ft. Distribution discontinuous and patchy, perhaps because of competition from spider monkeys. Primarily eats ripe fruit, but also leaves and other vegetation. Fur moderately long and very dense. Sexes similar in length, but males often heavier. Diurnal. Polygamous, in groups of 15–20 with several males. Gestation: 225 days. Litter size: 1. Births every 1.5–2 years, not obviously seasonal.

Humboldt's woolly monkey [V]

Lagothrix lagotricha
Humboldt's or Smokey or Common woolly monkey.

From N Colombia throughout Amazonia E to Rio Negro (N of Amazon) and Rio Tapajós (S of Amazon, S to N Bolivia. HBL 15–23in (female), 16–22in (male); TL 24–29in (female), 22–28in (male); wt 12lb (female), 14lb (male). Coat: body gray to olive-brown, to dark brown to black; head often darker, almost black.

Yellow-tailed woolly monkey [E]

Lagothrix flavicauda

Peru, in departments of Amazonas, San Martín, and La Libertad, in mid-elevation forest (5,500–8,900ft). HBL 20in (female); TL 22in (male); wt ? Coat: medium brown except for yellow underside to tail at tip around naked prehensile portion; fur on nose and mouth whitish.

Exploration and Play

Steps to adulthood in the Squirrel monkey

Squirrel monkeys are small, agile and inquisitive New World monkeys that live in thickets, mangrove swamps, and the lower layers and edges of tall forests. They are omnivores that feed mostly on fruit and insects, relying on their inquisitiveness and speed to catch the insects that are their preferred foods. Group sizes of 10 to 35 are common in smaller forests. In large expanses of Amazonian forests, groups may contain several hundred individuals. Adult females with their young travel together and form the core of the group. Adult males intermingle in the group during the several months of mating season, then take more peripheral positions during the remainder of the year. Squirrel monkeys are not territorial and home ranges often overlap extensively; however most groups avoid close contact with neighboring groups. Range size varies—the limited information available suggests day ranges of 37–62 acres (15–25 hectares), possibly up to 150 acres (60 hectares) or even 320 acres (130 hectares) in some areas.

In remote forests that are seldom visited by humans, Squirrel monkeys will appear out of the foliage to peer down at human visitors, cocking their heads and moving from branch to branch for better vantage points. The youngsters usually show the greatest interest and remain the longest. If an unusual bird perches in the trees, nearby monkeys may come up for a closer look. If all appears to be safe, the monkeys may even try to touch or catch the bird. Some birds appear to respond playfully to the monkeys' curiosity, flying just out of the monkeys' reach and waiting for the monkeys to approach again before flying just a few more feet away.

Young infants ride their mothers' backs for the first month of life, then begin venturing off to explore and play for increasing numbers of hours each day.

Both exploration and play involve physical exercise that promotes healthy physiological development, and both propel the young monkey into interaction with many aspects of its physical and social environments, where it learns information and skills that it will need as an adult to survive and to reproduce. During exploration of the physical environment, the young Squirrel monkey learns how to catch flying insects and other prey, open various types of foods, and move safely through the trees. It learns the location of food sources, water holes, wasp nests, and safe arboreal pathways.

Play involves higher levels of activity than does exploration. Active play in the trees gives the young monkey a chance to develop skills for running and jumping through the branches. These skills may be of critical importance if the monkey ever has to flee from a predator (eg a large hawk) or respond quickly in an emergency (eg when a branch breaks under its weight).

Social play helps the young develop bonds with their peers and become integrated members of their group. Various sexual, maternal and aggressive behaviors are practiced and perfected during social play; and this presumably facilitates the acquisition of reproductive skills and appropriate sex and gender roles. During social play, the young gain experience in transmitting and interpreting various communicative signals, such as facial expressions, body postures, intention movements and vocalizations. They often playfully imitate and acquire troop tradition, such as learning the location of trees with water holes and how to dip their hands to obtain water. Most social play involves wrestling, chasing and other forms of play fighting, in which individuals learn skills of sparring, dodging, jumping, pouncing, feinting, chasing and fleeing. Mastery of these self-defense skills can increase an individual's success in coping with predators and social conflicts, such as the fights that are common among adult males in the mating season.

Play among peers also functions to draw the infant away from its mother and help it become increasingly independent. Because play is rewarding and "fun," the maturing infants spend increasing amounts of time with peers, which eases the weaning process. By the end of its first year of life, the young Squirrel monkey spends hours each day with age-mates, almost completely independent of its mother.

In most primates (except humans) play ends by adulthood, though exploration continues in abated form for much longer. By adulthood, the monkeys have learned most of the information and skill that they need for survival; thus the benefits of continued exploration and play cease to outweigh the dangers of rowdy activity in the trees and the costs of high energy expenditure. Once the maturing monkeys have explored all the readily available parts of their home range and tried almost every type of game, they turn increasing attention to other activities. These may include foraging, sexual interactions, care of infants, dominance relations, cuddling together, and other "adult" activities, in which behavior is employed that has been acquired during exploration and play.

JDB

▲ **Juvenile Squirrel monkeys at play.** (1) Wrestling. (2) Sparring. (3) Exploring hole in tree. (4) Stalking a perched bird. (5) Learning to catch insect prey. Both social and solitary "play" are important in the apprenticeship for adult life.

► **Enterprise and curiosity** of the young Squirrel monkey.

Howling by the Light of the Moon

Why a "day monkey" has become the Night monkey

With the full moon overhead, a small monkey hoots mournfully in the treetops of the Amazon jungle. This is a lone male Night monkey searching for a mate; if his nightly travels of up to 3.7mi (6km) are unsuccessful, he must retire and try again on another moonlit night.

The Night monkey is the only truly nocturnal monkey and inhabits the forests of much of South America, where it feeds mainly on fruits, insects, nectar and leaves—with the occasional lizard, frog or egg. Night monkeys live in small groups of a male, female and young (a single offspring born each year and remaining with its family for two and a half years), occupying territories of up to 25 acres (10 hectares).

Unlike other strictly monogamous primates, only roving subadult male Night monkeys searching for a mate or adult males holding a territory will call, and then only when the moon is full, or nearly so, and the sky is cloudless. Female Night monkeys call rarely, if at all, in the wild. Similar groups of gibbons, siamangs and titis, for example, have daily morning duets to reaffirm territorial possession, males and females calling in unison, but each sex with a different song. Night monkeys do not sing in duets, and calling sessions are restricted to once or twice a month. On a clear night with a full or nearly full moon the adult male will give a series of 2–4 short, low hoots (10–30 hoots a minute) which can be heard for 550 yards (500m). The hooting monkey travels 330–1,150ft (100–350m) along or up to its territorial border during a 1–2 hour period, announcing his territorial possession. Night monkeys rarely fight during hoot nights.

But Night monkeys do fight. During a 12-month period the author observed 15 battles (about one each month), all occurring when the moon was bright and overhead, and invariably when a neighboring family group trespassed into a ripe fruit tree near a border. Males and females of each group burst into a low, ascending resonating "war whoop" and attacked. The home team won every time, putting the invaders to flight within 25 minutes. The three times during the year when groups met on a dark night, 5–10 short hoots were exchanged and the monkeys moved apart without fighting.

Why should bright moonlight be important to Night monkeys? Night monkeys can see well at low light levels with their enlarged eyes. They make spectacular ten- to 16-ft (3–5m) leaps from tree to tree, adeptly catch insects and locate fruit trees in light levels too low for humans or other diurnal primates to see. Yet even Night monkeys' activity seems to be limited in total darkness. Path lengths average about 1,800ft (550m) on dark nights, as opposed to 2,800ft (850m) on clear moonlit nights. Monkeys never fight or hoot extensively when there is no moonlight, and even rough-and-tumble play is restricted to dawn and dusk on moonless nights.

Night monkeys differ from most nocturnal mammals by having color vision, and the structure of the eye suggests that the ancestor of *Aotus* was active in daytime only. Why then, has a day monkey evolved into a night monkey? Other small South American monkeys, such as titis and marmosets, are hunted by diurnal hawks and eagles such as the Harpy and Crested eagles, and large monkeys, especially capuchins, chase smaller monkeys from fruit trees. The Night monkey avoids these two problems by sleeping, spending each day in the same

▲ **A monogamous family group,** with parents and single young of two annual birth seasons. Color patterning on the head is distinctive, as in many cebid monkeys.

◀ **More owl-like** than this parrot-like pose suggests, the Night or Owl monkey has a low resonant hooting call. Only males call, and then only on moonlit nights.

dense vine tangle or tree hole, only venturing out punctually 15 minutes after sunset, when daytime predators and competitors have roosted. The only nighttime predator big enough to eat a Night monkey is the Great horned owl, which is rare in the tropical rain forest.

In some habitats, however, such as the open forests of the dry *chaco* of Paraguay and Argentina, where diurnal Harpy and Crested eagles are rare, Great horned owls, common and capuchin monkeys absent, Night monkeys have partially reverted to daytime activity. They sleep on open branches (one group was observed to use 42 different sleep trees in five months) and are active for between one and three hours in daylight, feeding on fruits and flowers. In fact, in the cold *chaco* winter, during times of the month when there is no moonlight, groups travel nearly as far in the day (920ft/280m) as in the night (1,080ft/330m).

Although changes in ecological conditions modify certain Night monkey behavior, such as sleep, site selection and activity times, much behavior, such as hooting and fighting in the bright moonlight, do not change and are typical of the Night monkey in all habitats. PCW

Titi Monkey Family Life

Father is the primary caretaker of the young

Not all primates live in large social groups. Some, such as the indri, gibbons, siamang, Night monkey, the Mentawi Islands langur, and the three South American species of titi monkey live in small monogamous family units comprising an adult male, an adult female and their immature offspring. Elusive and difficult to observe, titi monkeys (genus *Callicebus*) are unique for the tail-twining posture which family members adopt when at rest, and are particularly notable for the prominent role played by the father in caring for the young.

There are three species of *Callicebus* monkeys. The Dusky titi (*C. moloch*) is found throughout most tropical forests of the Amazon, Orinoco, and Upper Paraná river drainage systems. The Yellow-handed titi (*C. torquatus*) frequently occurs in the same areas, but is found primarily on vegetation growing on white sands drained by black water rivers in Colombia, Venezuela and Peru and along the Rio Negro in Brazil. A third species, the Masked titi (*C. personatus*) is restricted to the dwindling Atlantic coastal forests of eastern Brazil in Bahia, Espírito Santo, Minas Gerais, Rio de Janeiro and São Paulo. All three species are small monkeys of about 2.2lb (1kg) but they differ substantially in coat color and pattern. For example, the hands of the Dusky titi vary from gray to brown to red, while the Masked titi has black hands and the Yellow handed titi, yellow hands. All titi monkeys live in family groups of 2–5 animals, occupy small territories, and regulate intergroup spacing with loud morning calls. About three-quarters of their diet consists of fruits, which they supplement with leaves, flowers and insects. The Dusky and Masked titis appear to obtain most of their protein from young leaves, leaf petioles and young shoots, and usually spend the last few hours of every day feeding on these food items relatively close to the ground. On the other hand, in vegetation growing on white sand soils the leaves are heavily laden with secondary (toxic) compounds, sclerophylous (hard and tough), and difficult to digest. The Yellow-handed titi therefore obtains most of its protein from insects, searching for them high in the forest canopy. The reduced feeding time required appears to allow an increase in the amount of time spent grooming.

Apart from territorial calls at dawn, titi monkeys move virtually noiselessly through the dense jungle except for an occasional soft "swoosh" and the patter of falling water droplets as they jump into the damp, pliable foliage of a neighboring branch. In this way

◀ **Typical social behavior** of titi monkeys. (1) Father carrying young. (2) Father grooming young. (3) Tail-twining when at rest huddled on a "sleep bough."

▶ **A Dusky titi** takes a meal of young vine leaves. Fruit is the major food of both Dusky and Yellow-handed species, but the Dusky titi may eat substantial amounts of young leaves and shoots, while the Yellow-handed titi takes insects instead.

▼ **Dusky titi with young.**

they hide from the huge Harpy eagle, or the voracious Ornate hawk-eagle.

Often during the day an adult pair stops, rests and grooms each other, usually with tails twined. The Yellow-handed titi has an additional period of extensive grooming just prior to sleeping: the entire group rests on a large "sleep bough," where all members of the family take turns grooming each other, with most time being devoted to grooming the youngest infant, particularly by the father. Finally, as the sun sets, the entire group huddles together to sleep with all their tails entwined. It has been suggested that balance and warmth are enhanced by tail-twining, but the primary function is probably social.

A striking feature of the behavior of titi monkeys is their long early morning bouts of calling which echo throughout the forest for distances of over a third of a mile (600–700m). Despite similarity in vocalizations, the species *Callicebus* do not all regulate intergroup spacing with the same vocal mechanisms. Dusky titi groups move to their territorial boundaries and confront each other in vigorous vocal battles. Boundaries are rigidly maintained over the years. In the Yellow-handed species, however, calling is more subtle, as groups do not confront each other at close range. They regulate intergroup spacing by calling from well within the territorial boundary, and over a period of years the boundaries shift. Nobody understands how or why these two patterns have evolved in two closely related species.

Titi monkey fathers are most indulgent and spend much time in the care, carrying and attentive observation of their offspring. Indeed infants have been observed to spend substantially more time in contact with their father than with the mother or siblings, and the father is the primary caretaker of the young. He will, for example, shift his position to cover and protect a young infant in a heavy rain storm. Whenever there is danger, as from strong winds or a falling tree, the father moves closer to his young. On one occasion when a juvenile animal fell to the ground, the father was seen to jump down quickly to within a few feet of the stunned animal to protect it until it was able to recover and move back into the safety of the trees. Most important, the father is responsible most of the time for carrying the infant through the trees—giving it up for the mother to nurse or occasionally for a sibling to carry it—until it is old enough to keep up with the adults on its own, at the age of 4–5 months.

A dawn call, which consists of repeated sequences of loud vocalizations, may last up to seven minutes. Neighboring groups do not call concurrently, but await the termination of the adjacent group's call before initiating their own. Usually an adult pair sits side-by-side and coordinates a vocal duet with each animal contributing a part of the song. If the pair moves apart the coordination breaks down and the male may continue singing alone. Young animals frequently join in with their parents for brief periods and one may even hear, for a few moments, a cacophony of four animals singing together. In this way a young animal begins to learn the species-specific song sequence. The completely coordinated duet is not fully learned until the animal grows up, leaves its natal family between two and three years of age, finds a mate, and establishes a new monogamous group. It takes up to a year for the newly formed pair to learn to coordinate its duet. WGK

Leaf-eaters of the New World

Diet and energy conservation in the Mantled howler monkey

The six species of howler monkey (genus *Allouatta*) are capable of producing calls that are among the loudest made by animals. Charles Darwin considered that natural selection favored those males which made the loudest calls to attract females. Others since Darwin have suggested that the howlers' calls are used to defend the troop's rights to particular food trees. In the case of howler monkeys, the latter explanation appears more correct, but it overlooks the probable energetic relationship between the development of loud howling behavior and the feeding patterns of howler monkeys, revealed in recent studies of the Mantled howler monkey of Central America and northwestern South America.

Howler monkeys have the widest geographical distribution of the New World primates; furthermore, where howlers occur, they usually make up the highest percentage of the local primate biomass. One factor contributing to the ecological success of howlers is their ability to use leaves as a major dietary component. Leaves are far more available and abundant in tropical forests than fruits, flowers or insects, so in such habitats leaf-eating primates are apparently faced with fewer problems of food availability than the more strictly frugivorous or insectivorous primates. Most primates, however, do not eat large quantities of leaves, despite their relative availability, because leaves are low in nutrients relative to indigestible fiber. Leaves are also very low in sugars—important energy sources for most primate species. To be a successful leaf-eater, a primate must therefore have some way of circumventing these problems.

In the tropical forests of the Old World, there are many leaf-eating primate species (eg the subfamily Colobinae). Colobine monkeys have specialized, sacculated stomachs similar in many respects to the complex stomachs of cows. Bacteria in the colobine stomach digest the fibrous cellulose and hemicelluloses in leaves. In this process (fermentation) energy-rich fatty acids are produced which, in turn, can be absorbed by the monkey and used to fuel its daily activities. It is only through the intervention of gut flora that monkeys or any mammals can utilize leaf fiber for energy.

Howler monkeys do not have a sacculated stomach but they do have two large sections in the hindgut (cecum and colon) where the necessary cellulolytic bacteria are found. In general, however, hindgut fermentors such as howler monkeys are probably not as efficient at obtaining nutrients and energy from leaves as the more specialized colobines, since in howlers fermentation takes place in sections of the gut below the small intestine—the major site of nutrient absorption. To improve digestive efficiency, howlers feed very selectively, primarily on tender young leaves that can be rapidly fermented and on a few species of unusually nutritious mature leaves. Howlers also eat sugary fruits and flowers whenever these are available but they can live for weeks at a time only on leaves,

▲ **Advertising its presence,** a male Mantled howler calls in warning to nearby troops. Such calls are thought to be one key to the success of howlers as a group.

► **About half of howlers' diet is leaves.** Adaptations in the stomach, a slow-moving life-style and careful selection of food also characterize howlers. This Mantled howler is feeding on leaves in the Panamanian forest; howlers are distributed also south as far as Argentina.

provided the leaves are high in quality.

Even with a careful feeding regime, howlers must still conserve energy, and they rely on behavioral and morphological adaptations to help. They are relatively slow-moving and more than 50 percent of the howler day is spent quietly resting or sleeping; during the day the monkeys range over only about 1,300ft (400m) and the home range for a troop of some 20 howler monkeys is just 77 acres (31 hectares). Thus all potential food sources within a howler's home range lie within an average day's travel time. Howlers also show a "division of labor" between the sexes; males help settle disputes within the troop and defend certain important food trees from neighboring howler troops, whereas females put their efforts into maintenance and reproduction and care of young.

The distinctive vocalization of howlers can be heard for well over a kilometer in their natural forest habitat. The howl is produced by passing air through the cavity within an enlarged bone in the throat, the hyoid, which is much larger in males. Howling itself contributes to economizing on energy; every morning around sunrise, each troop gives a "dawn chorus" that is answered by all other howler troops within earshot. A troop does not maintain an exclusive territory but shares part of its home range with other howler troops. By howling loudly each morning and again whenever it moves on during the day, one troop can inform another of its precise location. When two troops meet, as they occasionally do, there is a considerable uproar, with animals, particularly males, expending much energy in howling, running and even fighting. Thus it pays to avoid meeting another troop; howling is far less expensive in terms of energy expenditure than is patrolling the home range and looking for other howler troops or getting into long intertroop squabbles over food trees. There is a dominance hierarchy between troops and by listening to the various howls, weaker troops know the locations of stronger troops and can avoid meeting them during the day. This helps troops to space themselves more efficiently in terms of exploiting food sources. Thus, through a combination of adaptations in diet-related morphology and in spacing behavior, howler monkeys have surmounted problems that are usually associated with having leaves as a principal food source, and evolved into highly successful leaf-eating primates. KM

GUENONS, MACAQUES AND BABOONS

Subfamily Cercopithecinae
Forty-five species in 8 genera.
Family: Cercopithecidae.
Distribution: throughout Asia except high latitudes, including N Japan and Tibet; Africa S of about 15° N (Barbary macaque in N Africa).

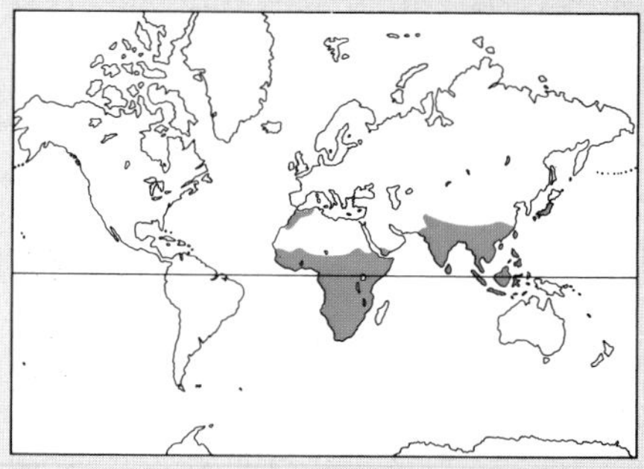

Habitat: from rain forests to mountains snowy in winter, also in savanna, brush.

Size: largest are the drill and mandrill, males of which (much larger than females) have a head-body length of 27.5in (70cm), tail length of 4.7 and 2.8in (12 and 7cm) and weight up to 110lb (50kg). Smallest is the talapoin, with head-body length 13.4–14.6in (34–37cm), tail length 14.2–15in (36–38cm) and weight 1.5–2.9lb (0.7–1.3kg).

Coat: long, dense, silky, often (especially in males) with mane or cape. Colors generally brighter in forest-dwelling species. Skin color on face and rump important in other species.

Gestation: 5–6 months.

Longevity: 20–31 years according to species.

Macaques (*Macaca*, 15 species), including **Rhesus macaque** (*M. mulatta*), **Stump-tailed macaque** (*M. arctoides*), **Barbary macaque** [V] (*M. sylvanus*), **Bonnet macaque** (*M. radiata*), **Toque macaque** (*M. sinica*), **Japanese macaque** (*M. fuscata*), **Lion-tailed macaque** [E] (*M. silenus*), **Père David's macaque** (*M. thibetana*), **Crab-eating macaque** (*M. fascicularis*).

Baboons (*Papio*, 5 species), including **Savanna baboon** (*P. cynocephalus*), **Hamadryas baboon** (*P. hamadryas*), **drill** [E] (*P. leucophaeus*), **mandrill** (*P. sphinx*).
Gelada baboon (*Theropithecus gelada*).

Mangabeys (*Cercocebus*, 4 species), including **Gray-cheeked mangabey** (*C. albigena*) and **Agile mangabey** (*C. galeritus*).

Guenons (*Cercopithecus*, 17 species), including **Vervet** or Grivet or Green monkey (*C. aethiops*), **Blue** or **Sykes' monkey** (*C. mitis*), **Redtail monkey** (*C. ascanius*) or Schmidt's guenon, **Mona monkey** (*C. mona*), **De Brazza's monkey** (*C. neglectus*), **L'Hoest's monkey** (*C. l'hoesti*).

Patas monkey (*Erythrocebus patas*).
Talapoin monkey (*Miopithecus talapoin*).
Allen's swamp monkey (*Allenopithecus nigroviridis*).

[E] Endangered. [V] Vulnerable.

Tough, active and gregarious, noisy, imitative, curious—the cercopithecines are the "typical monkeys" best known to legend because their distribution and ways of life bring them into contact with people. Many are adaptable generalists, able to take advantage of the wastefulness or sentimentality of human neighbors to make a living, or to take their share from unwilling human hosts by skillful theft from unharvested crops or food stores.

Previously, when a laboratory scientist referred to "the monkey" he nearly always meant the Rhesus macaque, a cercopithecine which has long borne the brunt of our invasive curiosity.

It is only in recent years that we have begun to appreciate the variety within the subfamily Cercopithecinae. "The monkey" was a useful concept to those looking for a non-human model of human disease, or those who saw the animal kingdom as a static hierarchy with man at the top of the ladder. But the modern comparative approach towards natural history, made possible as we learn more about the anatomy, physiology and behavior of different species, is more exciting: it includes insight into the dynamics of the group, and the possibility of recent, ongoing, and future change in this young and probably most rapidly evolving subfamily of primates.

Anatomy and posture of the "typical monkey"

Cercopithecine monkeys have the same dental formula as man: I2/2, C1/1, P2/2, M3/3 = 32. They have powerful jaws with the muscles arranged to give an effective "nutcracker" action between the back teeth. The face is rather long, except in some of the smaller guenons, the "dog-faced" baboons being the most extreme in this

▲ **Foraging in grass,** a female Olive baboon in Kenya bears her watchful infant on her back. Baboons typically stand on one hand and pluck grass with the other.

◄ **The vervet or grivet** ABOVE LEFT is the most widespread of the guenons, and lives in many local variants throughout African savanna. Other guenons are mostly forest dwellers.

◄ **The White mangabey** is a medium-sized monkey whose large incisor teeth enable it to tackle food (eg palm nuts) that cannot be used by guenons that share the same forest. The red-capped race occurs in Cameroon.

trend: the longer jaw increases the surface area of back teeth available for grinding food. The canine teeth are much longer than the rest in all males, and in female guenons, but not in adult female baboons, mangabeys and macaques. Long upper canines cut against a specially modified "sectorial" premolar in the lower jaw.

Cercopithecines have cheek pouches which open beside the lower teeth and extend down the side of the neck. When both are fully distended, the cheek pouches contain about the equivalent of a stomach's load of food. When competing for food, or foraging in a dangerous place, they cram it hastily, with minimal chewing, into the pouches and retire to a safe place to eat at leisure, leaving the hands and feet free for running and climbing. In a gesture highly characteristic of cercopithecines, the back of the hand is used to push food in the full pouches towards the opening. Food is processed (peeled, scooped out, etc) by hand and mouth, under close visual scrutiny, so that what is swallowed is selected for high quality and digestibility. The stomach is simple and not very large, and the gut as a whole is unspecialized.

The first digit can be opposed on both the hands and feet, a facility used equally in moving, feeding, and such social activity as grooming. The hind foot is plantigrade (the animal walks on the whole foot, not just the toes) in all cercopithecines, and the hand also is plantigrade in primarily arboreal species. In those other species which move from place to place primarily on the ground, such as baboons and the Patas monkey, the joint between palm and fingers forms an extension of the walking limb, so that the walking surface of the hand is the underside of the four fingers, the palm being held vertically.

In tree-living species the hindlegs are long and well-muscled, used in leaping between branches and bounding along them. When moving slowly in the trees the hind legs are often forward under the body so that the center of gravity is over a line from knees to feet and the hands are almost freed from weight-bearing and are available for manipulation. This posture is especially characteristic of species with a long heavy tail which helps to bring the center of gravity back over the hips. On the ground, aboreal monkeys are "down at the front"—their arms being shorter than their legs. Monkeys which spend more time on the ground have relatively longer front limbs. The most specialized ground walkers, the baboons, have their shoulders higher than their hips when walking, due to the lengthening of the arm bones and the addition of the palm to the effective limb length. The smaller Patas monkey has not achieved the "shoulders higher" posture, so, in order to scan its surroundings, it frequently stands upright.

The stance of a baboon requires support under the shoulders. When foraging, therefore, baboons typically stand on one hand and pluck grass with the other. Gelada baboons typically sit to forage with both hands, before shuffling forward on their bottoms. Ground-walking monkeys do not require tails for counterbalancing, and many have shortened tails or none at all—the absence of a tail does not reliably distinguish apes from monkeys as is popularly supposed (and suggested by the name cercopithecine, "tailed ape"). Monkey tails seem to have very poor circulation and are very susceptible to frostbite, so a shortened tail can be an adaptation to a cold climate, although some short-tailed monkeys live in warm places. Both short and long tails are used to communicate: forest monkeys hold their tail in a position characteristic of the species, but in general the long tails of arboreal monkeys are of limited use in communication, because of the requirements of balance and locomotion. The tails of terrestrial monkeys, on the other hand, are sometimes used to signal the mood of the owner, particularly confidence or fear. Patas females curl up their tails when they are sexually receptive. The first few vertebrae in the tail of the Savanna baboon fuse during the first year to form an upright stalk from which the rest of the tail dangles, at an angle that steepens gradually as the baboon matures.

Coat, changes in skin color and swellings

Cercopithecines have long, dense, rather silky fur, often with longer hair forming a cape or mane over the shoulders which is heavier in adult males. The macaques, baboons and mangabeys have drab brown or gray coats—a few macaques are black. Most guenons have brightly colored coats with vividly contrasting patterns, especially around the face. These colors make the forest-dwelling species easy to identify, even

▼ **Drills and baboons,** largest of the monkeys (adult males shown). (1) Mandrill (*Papio sphinx*). (2) Drill (*P. leucophaeus*). (3) Gelada (*Theropithecus gelada*), showing bare patches on neck and chest. (4) Hamadryas baboon (*Papio hamadryas*) with red naked skin on face and rump. (5) Guinea baboon (*P. papio*). Three forms of the Savanna or Common baboon (*Papio cynocephalus*): (6) Yellow baboon of lowland East/Central Africa asleep in a tree; (7) Olive baboon of highland East Africa with hare prey; (8) Chacma baboon of southern Africa.

3

2

1

where several live together. In baboons, mangabeys and macaques the bright colors occur on patches of bare skin on the face, rump, and, in the gelada, the chest. The red face color of some macaques depends on exposure to the sun: monkeys in the wild have red faces, those kept indoors have gray-white ones. The black face color of the Patas monkey also fades indoors. Skin color also depends on gonadal hormones, so that adults but not juveniles show the bright red; the color is brighter still during the mating season, and castration causes the color to fade. Mandrills have patches of blue skin as well as red, but the blue is a structural, not a pigment, color. A similar bright blue color appears in the scrotal skin of several guenons at adolescence. This requires gonadal hormone (testosterone), but once formed it is stable, and does not disappear if the male is castrated.

Female baboons, mangabeys, and some macaques and some guenons (see table of species, pp88–89) develop a perineal swelling which in normal animals increases in size during the first half, or follicular, phase of the menstrual cycle and decreases after ovulation. The swellings of adult females clearly identify a female which is about to ovulate, and adult male baboons, for example, will not normally attempt to copulate with a female unless she has a swelling. The exact site of the swelling differs from species to species, and each individual also has recognizable swelling patterns. The swellings tend to increase in size in successive cycles, so that captive monkeys which are not breeding regularly may grow enormous swellings, up to 10 or 15 percent of their body weight. In the wild, adult females rarely undergo more than one or two successive cycles before conceiving

6

4

5

7

8

again, so have smaller swellings. Rhesus macaques and some other species show a rather different pattern of swelling, associated with adolescence and sometimes, in the adult, with a return to reproductive activity after a long interval. The adolescent Rhesus female swells around her tail root and along her thighs, the swelling reaching a maximum just before menstruation, in the luteal phase of her cycle. This pattern ends when she is fully mature and ready to conceive. Similarly, adolescent Patas females show vulval swelling, whereas adults usually do not.

In the wild, changes which occur during pregnancy are even more conspicuous in some species than these changes related to the menstrual cycle. Thus the black naked skin over the ischia on the rump of baboons loses its pigment and becomes a glowing red in pregnancy. The vulva of Vervet monkeys and mangabeys also reddens when they are pregnant, sufficiently to be easily recognized at a distance.

Newborn cercopithecines have short velvety fur, often of a color that contrasts with the pelage of the adult: olive-brown Stump-tailed macaques have primrose-yellow infants, while yellow-gray baboons have black infants. The infants are even more conspicuous because they have little or no pigment on the naked skin of face, feet, perineum, and ears (which seem to be almost adult sized). The newborn male baboon has a bright red penis, while infant male macaques have a very large, empty scrotum. Infant guenons, on the other hand, are very difficult to sex even when you have them in your hand. Special attention to infants by other than their mother often coincides with the duration of the natal coat; such attention includes frequent inspection of the newborn's genitals, and it is perhaps important for the normal development of social behavior that the sex of the infant is known to its fellow group members as early as possible. The natal coat begins to be replaced in the third month by a juvenile coat which is usually a fluffier, less brightly colored or clearly marked version of the adult pattern.

Evolution

The cercopithecines are a modern group which originated in Africa, their macaque-like forebears appearing in the fossil record towards the end of the Miocene (26–7 million years ago). Macaques spread north and east, to Europe in the Pliocene (7–2 million years ago), and Asia by the Plio-Pleistocene (2 million years ago). At the same time they died out in Africa south of the Sahara, or perhaps evolved and radiated into the modern African genera, the baboons, mangabeys and guenons. The Barbary ape seems to be somewhat separated from the Asian macaques in terms of molecular genetics as well as some anatomical details, perhaps retaining some more primitive features.

The radiation of the guenons in Africa is very recent: the emergence of distinct species (speciation) seems to have occurred during the last main glaciation, when Africa as a whole was colder and much drier, so that the forests retreated to a few scattered areas in equatorial Africa—along the east and west coasts, and around Mount Cameroon and the Ruwenzoris. When the climate became wetter and warmer about 12,000 years ago the forests spread, and with them their now distinct monkeys. Similarly, in Asia, cold dry periods associated with glaciation restricted the habitats of macaques and separated populations, which differentiated. Here the picture is more complicated because variations in sea level allowed macaques to move from one island of Southeast Asia to another.

In Africa, the history of baboon species is the history of grasslands. The earliest baboons were ancestors of today's gelada, specialized to harvest grass seed. Dry grassland spread in dry periods at the expense of forest, encouraged by human habits of burning to control game and later to increase grazing for livestock. The Savanna or "Common" baboon, once probably a forest-edge species, spread with the wooded savannas which resulted, and eventually replaced the geladas except in their Ethiopian highland stronghold.

Species and distribution

All **macaques** are placed in a single genus (*Macaca*) which occupies the whole of Asia except the high latitudes. Most areas are occupied by a single species with characteristics appropriate to local conditions. The one surviving macaque in Africa, the Barbary ape, has thick fur and no tail, traits which help it to survive the snowy winters of the Atlas mountains where it lives. Another shaggy and short-tailed species, the Japanese macaque, survives the snowy winters of northern Japan, and yet another, Père David's macaque, lives in high mountains of Tibet. The sturdily built, medium-tailed Rhesus monkey of the Himalayan foothills, northern India and Pakistan (see p90), is replaced in southern India by the smaller, more lightly built and longer tailed

3

▲ **Facial expressions in macaques** (adult males shown). (1) Barbary macaque (*Macaca sylvanus*): "lip-smacking" with infant: see p92. (2) Moor macaque (*M. maura*): open-mouth threat. (3) Bonnet macaque (*M. radiata*): canine display yawn. (4) Crab-eating macaque (*M. fascicularis*); fear grin. (5) Stump-tailed macaque (*M. arctoides*): open-mouth threat. (6) Pig-tailed macaque (*M. nemestrina*); approach pout-face – can precede copulation or attack or even grooming another individual. (7) Rhesus macaque (*M. mulatta*); aggressive stare.

Bonnet macaque which, further south still in Sri Lanka, is in turn replaced by the rather similar Toque macaque. Towards the Equator, in wetter areas which will support rain forest, the number of niches available for monkeys increases, and two species of macaque occur together. Thus the arboreal Lion-tailed macaque lives in the forest of southwest India above the more terrestrial Bonnet macaque. In Sumatra and Borneo the small, long-tailed, Crab-eating macaque lives in the same forests as the large, heavily built, short-tailed, more ground-living Pig-tailed macaque. In Sulawesi (Celebes) a black, stump-tailed group of species has radiated (3, 4 or 7 of them according to different authors) to occupy the different peninsulas of that dissected island, with only one species in any given locality. The Stump-tailed macaque is another short-tailed species which inhabits the high mountains of Southeast Asia, while other macaques occupy the rain forest below.

The **baboons** live everywhere in Africa where they can find drinking water. They have dog-like muzzles, and limb modifications which allow them to walk long distances on the ground. The Savanna or "Common" baboon is the most widespread; individuals from different regions are sufficiently different in appearance to have been given separate species status in the past. They live in grass- and bush-land, and along the edge of forests. The Hamadryas baboon replaces the Savanna baboon in the Ethiopian highlands: male Hamadryas baboons have red faces and rumps instead of black, and a long cape of gray fur, so they look very different (though females are similar to females of the West African Guinea baboon). The two species hybridize along a narrow boundary zone in Ethiopia. The drill and the mandrill are baboons of the forest floor of west central Africa. They have contiguous distributions in Cameroon occupying the forest north and south of the Sanaga River respectively. The largest of the baboons, they are mainly black and have short tails. Their niche is perhaps parallel to that of the Pig-tailed monkey, largest of the macaques. The gelada is a long-haired species living in the highlands of Ethiopia, and specialized in gathering grass seeds.

The **mangabeys** (*Cercocebus* species) may

be thought of as lightly built, long-tailed baboons. They only live in closed-canopy forests. Some, like the Gray-cheeked mangabey, are highly arboreal, while others, like the Agile mangabey, usually move on the forest floor. An arboreal and a terrestrial species may live in the same forest: the author has seen Gray-cheeked and Agile mangabeys in the forest reserve of Dja, southern Cameroon.

The **guenons** are mainly in the genus *Cercopithecus*, but a few of the more eccentric species have been given generic status. The many species of *Cercopithecus*, recognized by differences in coat color, are regional variations of perhaps half a dozen ecotypes. In a given forest one finds only one example of an ecotype, so that in the richest habitats there may be four or five guenon species living together. The Vervet, Grivet, or Green monkey is the most widespread guenon, living in many local variants throughout savanna Africa. It is never far from water, typically inhabiting the acacia trees which grow along water courses. All the other species of *Cercopithecus* are forest species. The most widespread is the Sykes' and Blue monkey (*nictitans-mitis*) group, large guenons which include quite a lot of leaves in their diet and are found wherever there is a patch of closed-canopy forest; next come the red-tailed guenons (*cephus-ascanius* group), smaller monkeys which seem to require a more layered canopy to the forest, with tangles of creepers; then the *mona* group, smaller still, and more insectivorous. These three species commonly form associations of monkeys of more than one species, in rain forests of Gabon for example. De Brazza's monkey inhabits wet patches of forest which include palms. The ground-living L'Hoest's monkey can inhabit quite high-altitude forest. Where overlap occurs, some hybridization of species is known (see p100).

Other guenon genera have only a single species each. The Patas monkey is accorded generic status because of its skeletal adaptations to ground living. The largest of the guenons, it is like a long-legged, orange-colored vervet. It lives in the open acacia woodlands and scrub of drier more seasonal areas north of the equatorial forest. Its habitat is often adjacent to that of the vervets and, although much larger, patas avoid vervets when they meet. The talapoin is the smallest of all the Old World monkeys, and lives in floodplain forests of west central Africa; Allen's swamp monkey is another swamp-living monkey from further east, in the Congo basin. Both these are placed in separate genera partly because the females have a perineal swelling, unlike females of other guenons.

Diet and foraging

Cercopithecines are primarily fruit-eaters, but their diet may include seeds, flowers, buds, leaves, bark, gum, roots, bulbs and rhizomes, insects, snails, crabs, fish, lizards, birds, and mammals. They take almost anything which is digestible and not actually poisonous. Most food is caught or gathered with the hands. Selection and preparation of food is learned from observation, initially of the mother. In this way local traditions in food preference develop. Adult baboons will prevent juveniles from eating unfamiliar food. On the other hand, in experimental situations, juveniles have come to recognize new food items and devised new preparation methods; other juveniles and adult females have learned from them, but adult males less readily so. This transmission of information may be the most important foraging-related function of group-living: the troop is primarily an educational establishment.

Species which live near water use aquatic foods. Japanese macaque troops living by the seashore have recently incorporated some seaweeds into their diet. The Crab-eating macaque is so called for good reason; Savanna baboons at the coast in South Africa take shellfish off the rocks; and talapoins are said to dive for fish.

Where several species live in association, for example in a West African forest, the smaller species tend to eat more insects, and

▲ **A Crab-eating macaque** in its typical riverbank habitat. Crabs may indeed be taken by this omnivorous species found from Indonesia and the Philippines to Burma.

► **The Patas monkey** RIGHT ABOVE is larger, more long-legged and more terrestrial than its forest guenon relatives. Inhabiting the open acacia woodlands north of Africa's equatorial forests, it runs from one tree to the next in search of the fruit, galls, leaves, flowers and gums that comprise most of its diet.

► **"Red, white and blue" display** of the male Vervet monkey. The Vervet monkey is closely related to the Patas monkey and takes a varied diet including fruit, leaves, flowers, insects, eggs, nestlings, rodents and crops. Bright blue scrotum and red penis are characteristics of both Patas and Vervet monkey males.

▼ **Savanna baboon eating gazelle kid.** Baboons may act cooperatively to head off and trap such prey.

more active types of insects (like grasshoppers), while larger species eat more caterpillars, and more leaves and gum. Mangabeys have powerful incisors, used to open hard nuts which are inaccessible to guenons.

Patas monkeys are adapted for running in grassland between patches of the acacia woods where they feed on fruit, leaves and gum, as well as insects and some small vertebrates. Their diet is not very different from that of other guenons living in forests. Baboons, in contrast, have a diet that includes large quantities of grass, and the large area of molar teeth, made available by the longer jaws, allows them to prepare the tough silicaceous leaves. Their very powerful spoon-shaped hands are strong enough for digging, and they subsist through severe dry seasons by digging up the rhizomes and the leaf-base storage bulbs of grasses and lilies. The small incisors of geladas are suitable for extracting seeds from grass heads, and they are also "close grazers," using a rapid pinching movement of the thumb and forefinger of each arm alternately to crop a sward as closely as sheep. Baboons take small mammalian herbivores, kill and eat the kids of gazelles which are left hidden in the grass, and hunt hares which they start from their forms. In a simple form of cooperative hunting, a group of baboons will spread out to start and head off such small prey, almost like beaters, although the prey is shared only reluctantly by the eventual captor.

Wherever people and monkeys come in contact (which is relatively frequent, as their ecological requirements are rather similar) the diet of the monkeys expands to include offerings, garbage, or stolen crops. The monkeys' behavior is clear evidence that learning plays an important role in how they acquire food. They time their arrival at feeding stations to coincide with the arrival of food, and raid crops when people are predictably absent, in heavy rainstorms or during the siesta. Baboons will enter a field where women are working, and even chase them away, but avoid men, who are usually armed. Talapoins will crowd quite close to people who are washing or fishing at a river in the forest, but avoid people setting out to hunt; all of which suggest a sophisticated appreciation of human behavior.

Predators

Monkeys are themselves prey to other animals. Some of the largest eagles feed mainly on forest monkeys, for example the African Crowned hawk eagle which often hunts in pairs. One swoops and perches among a troop of monkeys, and while they mob it, the mate swoops from behind and picks up an unwary monkey. Forest monkeys use a special alarm call when an eagle flies over, and respond by diving into thick cover. The Crowned hawk eagle can, however, fly through forest and hunt on the forest floor, and has been seen to kill a near-adult male mandrill, largest of all the cercopithecines. In open country the Martial eagle may prey on vervets and baboons. Vervets are the prey of pythons, which wait for them at the base of trees. These monkeys make specific alarm calls on sight of poisonous tree snakes. Monkeys are probably only incidental food items for Carnivora. Apart from birds, other primates may be the most important predators of monkeys. Baboons occasionally take vervets at Amboseli in Kenya, and chimpanzees eat baboons, red colobus and guenons at Gombe in Tanzania. Monkey is also the preferred meat of some people in West and Central Africa and in Southeast Asia.

Mating and the raising of young

Cercopithecine monkeys are slow to mature, slow to reproduce, and live long. The Rhesus macaque usually conceives first at 3½ years of age and gives birth at 4, but perhaps 10 percent will mature a year earlier, and another 10 percent a year later than that. Patas, the largest guenons, usually conceive at 2½ years (range 1½–3½ years). They are thus the fastest-maturing cercopithecine monkeys so far recorded. On the other hand, females of the smallest cercopithecine, the talapoin, do not conceive until 5–6 years old, and the same is true of other forest guenons, such as the Sykes' monkey and De Brazza's monkey. In some species the age at first conception is variable, probably depending on nutrition. Captive baboons, for example, may conceive at 3½ years, but baboons in a deteriorated natural environment, as at Amboseli, not until 7½ years. Vervets at Amboseli conceive first at about 5 years, while in captivity they conceive at about 2½ years. This difference has not been observed in forest guenons or patas. Males begin to produce sperm at about the same age as the females of their species first conceive, but they are not then fully grown and are still socially immature. Males are several years older than females before they begin to breed.

Most cercopithecines conceive in a limited mating season. In high latitudes mating occurs in the fall. In the tropics, conceptions occur in the dry season among guenons in

◀ **Clinging to mother's belly,** this infant Vervet monkey will be nursed up to the arrival of the next single offspring, about a year after its own birth.

▼ **Bonnet macaque mother and young.** The mother/infant bond continues into adulthood with daughters, but sons leave the group on becoming sexually mature. Central whorl growing out sideways on crown gives the species its name.

wet forests, but in the wet season among patas in dry country. Baboons and mangabeys breed at any time of the year, but stress, such as a drought, that causes the death of several infants may have the effect of synchronizing the next pregnancies of the mothers. Other factors that determine mating periods include increased food supply and social facilitation. Mating periods may last for several months, as in vervets, or be concentrated into a few weeks, as in patas and talapoins, in which case the female probably ovulates only once in the annual season. Individual females usually mate in bouts lasting several days; during a species' mating period, females which have already conceived will continue to show bouts of receptivity and to copulate. Male Rhesus macaques also show seasonal changes, with reduced testosterone levels and testis size in the summer months.

Courtship is not elaborate, since mates are usually familiar with each other; signals indicating immediate readiness to mate generally suffice. The courtship of patas females is perhaps the most elaborate, a crouching run with tail tip curled, chin thrust forward, lips pouted. The patas female also puffs out her cheeks, and often holds her vulva with one hand or rubs it on a branch at the same time. Consortships are frequently formed, a pair remaining close together and at the edge of their troop for hours or days. In some species a single mount may lead to ejaculation, in others several mounts precede it. Copulating pairs are often harassed by juveniles, especially by the female's own male offspring, and the disruption may be enough to make repeated mounting necessary before ejaculation is achieved.

A single infant is born after a gestation of 5–6 months (twins are very rare), furred and with its eyes open; it often grasps at the mother's hair with its hands even before the legs and feet are born. Infants cling to the mother's belly immediately and can usually support their own weight, although the mother typically puts a hand to the infant's back, supporting it as she moves about during the first few hours. The newborn usually has the nipple in its mouth and uses it to support its head even when it is not nursing. Most monkeys are born at night, in the tree where the mother sleeps, and she eats the placenta and licks the infant clean before morning. (None of the cercopithecines ever makes a nest.) Patas infants, on the other hand, are usually born on the ground and during the daytime. It seems that timing of births is subject to selection by predation pressures, for Patas monkeys sleep at night in low trees, where they are vulnerable to predators. Changes in maternal care match the growing ability of the infant to move independently. Nursing becomes infrequent after the first few months, but usually continues until the next infant is born. The next birth usually occurs after about a year in most macaques, vervets and Patas monkeys and talapoins, and after two years or more for forest guenons like the Blue monkey. In baboons the birth interval varies, probably depending on food availability, between 15 and 24 months. If an infant is stillborn or dies, the birth interval may be shortened, although not significantly so in species that breed seasonally.

The mother–daughter bond lasts into adulthood, and the maternal bond with sons lasts until sexual maturity, when juvenile males of most species leave their natal group and enter another one or become solitary. Beyond infancy, the bond is seen in the frequency of grooming or sitting together, and in defense of both juvenile by mother and vice versa. Juveniles also form bonds with their siblings, and where hierarchies are in evidence, a female usually ranks just below her mother and just above her older sisters. Males lose their inherited rank when they leave the troop, but a young rhesus male will often join the same new troop as his older brother, who helps with his introduction.

When more than one female is receptive in a troop, there is some tendency for males to prefer older females, which also have rather longer periods of receptivity. Similarly females tend to prefer older males. Baboons may have preferred mates, spending time together even when the female is not receptive (see Olive baboon, p96). In species where a single adult male lives with a group of females (see below) other males come to the troop when several females are receptive, thus providing the females with a choice of mates.

Matrilines and troops

The basic unit of cercopithecine social oranization is the matriline, in which daughters stay with their mothers as long as they live, while males usually leave the natal group around the time of adolescence; the Hamadryas baboon seems to be the single exception (see p98). Whereas some years ago it seemed possible to identify species-typical group sizes, home ranges, and social organization, recent research has revealed considerable variation within a species, both in time and from place to place.

Under good conditions, a single founding female may be survived by several daughters, each now head of her own matriline, and all still within a single troop. In harsh conditions, survival may be so low that the matrilinear organization can only be detected after several years of study.

An upper limit to troop size may be determined by constraints on foraging; macaque troops that are provisioned with food begin to split into smaller troops, mainly along matrilines, but only after troop numbers have reached the hundreds, more than occurs naturally. Combination or fusion of troops is even rarer. In general, troops retain a defined home range; guenons defend a territory against adjacent troops. The range is the "property" of the females which form the permanent nucleus of the troop, and in Blue and Redtail monkeys (pp100–101) it is the females, with juveniles, which engage in boundary disputes. The male troop members are transitory. They may remain in a troop for a few weeks, or months, or 2–3 years (rarely more). Adult males make loud calls which are highly species-specific and (where they have been studied) also characteristic of an individual male (see pp94–95). The loud calls serve as a rallying point, and may also locate the troop and identify the male as being in residence. Thus the males' loud calls provide a means of communication between troops which are sub-units of the larger population, while the females tend to keep the troops apart.

Cercopithecines have been categorized into one-male and multi-male group species. Male baboons, mangabeys, and macaques will tolerate each other's presence in a troop; nonetheless, a small troop may still include only a single fully adult male. Males living together in a troop will establish a hierarchy based on the outcome of competitive interactions. The rank order is not very stable, but changes with age, or as males join or leave the troop. In some studies, males that ranked high in the male hierarchy also did most of the mating, while others have found no such correlation.

Among guenons, the vervets and talapoins also live in multi-male groups, but patas and those forest guenons that have been studied have one-male troops. This single male's tenure may be quite short and, in a mating season when more than one

3

◀ **Adult females** remain with their mothers but males leave the troop. Patas monkeys, like these at a water hole, and forest guenons live in one-male troops.

▼ **Small and medium-sized cercopithecines.** (1) Gray-cheeked mangabey (*Cercocebus albigena*), western race with double crest. (2) Allen's swamp monkey (*Allenopithecus nigroviridis*). (3) The Moustached monkey (*Cercopithecus cephus*) bobs its head from side to side when threatening, to "flash" his mustache. (4) Face of Sooty mangabey, a geographical race of the White mangabey (*Cercocebus torquatus*). (5) Talapoin (*Miopithecus talapoin*), the smallest Old World monkey. (6) (6) Patas monkey (*Erythrocebus patas*): long legs, long feet and strong, short digits are adaptations for running in this fastest-moving of all primates, which may attain 35mph (55km/h).

female is receptive, other adult males join the troop and also copulate.

Newly arrived adult males have been seen to kill infants they find in the troop, leading some observers to regard this as part of the male's reproductive strategy (see p114), though other studies of the same species have not revealed infanticide.

Adult male guenons not in a troop are usually found alone, although patas males will form small temporary parties. In captivity more than one male can be housed together only so long as no females are present. Talapoins live in very large multi-male troops, but outside the mating season the males live in a subgroup whose members interact with each other but very rarely with females. Hamadryas baboons (see p98) and geladas have "harem" groups within troops. Each adult male gelada herds several females, while bachelor males live in a peripheral subgroup. Baboons move in procession, usually with adult males at the front and rear, adult females also towards front and rear (including those carrying infants), and juveniles towards the center of the column.

Conservation

All forest-living monkeys may be considered endangered, because tropical forests are being destroyed at such a high rate and monkey populations are always at risk because of their slow reproductive rate. Where monkeys are considered a delicacy, the introduction of guns and the increased commercialization of hunting have further greatly reduced populations. As crop-growing areas are extended, the displaced monkeys raid crops: modern cash-oriented economies are less tolerant of such theft than are traditional societies. Monkeys also share many diseases with people—tuberculosis is one human disease to which monkeys are susceptible. Monkeys have been shown to carry yellow fever, to which they are very susceptible, and baboons carry asymptomatic schistosomiasis. There have been occasional suggestions that monkeys be exterminated to control disease, but probably no actual attempts to do so. For several years it seemed as if the increasing demand for monkeys for use in medical research, together with the appallingly high mortality rates in trapping and shipping, would cause the extinction of some "popular" species. Recently, a decline in the research industry, increasing efforts to breed monkeys in captivity for research purposes, and awareness of the need to handle newly caught animals carefully have reduced this threat. But the conservation of monkey species is fundamentally a matter of preserving the ecosystems in which they live, in large enough patches to allow viable populations to survive. Successful management depends upon controlling human encroachment. TER

5

6

THE 45 SPECIES OF "TYPICAL" MONKEYS

Abbreviations: HBL = head-and-body length; TL = tail length; wt = weight.
[E] Endangered. [V] Vulnerable.

Mangabeys

Genus *Cercocebus*

Medium-sized monkeys restricted to forests and closely related to the baboons. The brownish species (Agile and White mangabeys) are considered to be closely related to each other and rather widely separated from the blackish species (Gray-cheeked and Black mangabeys), for which the genus name *Lophocebus* has been proposed but not widely accepted. All have tails longer than their bodies. Females smaller than males but not as markedly as in guenons. Their large strong incisor teeth allow mangabeys to exploit hard seeds which are not accessible to guenons, with which they share habitats.

Pregnancy lasts about 6 months and there is no evidence of breeding seasonality. Infants are the same color as adults. Mangabeys live in large groups which include several males. They are very vocal, and the adult male has a dramatically loud long-distance call (the whoop-gobble of the Gray-cheeked mangabey, p94), while the adult females of a group also perform loud choruses.

Gray-cheeked mangabey

Cercocebus albigena

SW Cameroon to E Uganda. Primary moist, evergreen forest. Arboreal. Diet: fruit and seeds, also flowers, leaves, insects and occasional small vertebrates. HBL: males 18–24in, females 17–23in; wt males 20lb, females 14lb. Coat: black, with some brown in long shoulder hair; short hair on cheeks grayish; hair on head rises to single (eastern races) or double (western races) crest. Female has bright pink cyclic vulval swelling.

Black mangabey

Cercocebus aterrimus

Zaire. Rain forest. Arboreal. Diet: fruit, seeds. HBL 28in; wt about 22lb (male). Coat: black.

Agile mangabey

Cercocebus galeritus
Agile or Tana River mangabey.

Cameroon and Gabon, Kenya and Tanzania. The recently discovered eastern populations are scattered and separated from the western ones by thousands of miles. Rain forest. Terrestrial. Diet: palm nuts, seeds, leaves. HBL 17–23in; wt 12lb (female), 22lb (male). Coat: dull yellowish-brown; hair on top of head forms crest. (Includes *C. agilis*.)

White mangabey

Cercocebus torquatus
White, Collared, Red-capped, or Sooty mangabey.

Senegal to Gabon. Primary rain forest. Terrestrial. Diet: palm nuts, seeds, fruit, leaves. HBL 26in; wt about 22lb (male). Coat: gray; geographical races have color variants: a white collar in Ghana, a red cap in Cameroon. Females have a cyclic vulval swelling.

Guenons

Genus *Cercopithecus*

The African long-tailed monkeys which are mainly forest living. Both sexes have brightly colored coats, but with patterns more pronounced in males. Infants have dark or dull-colored coats, with pink faces at birth which darken later. Tails are considerably longer than bodies. Males are much larger than females, the difference being greater in larger species; females range from two-thirds to half weight of male. Taxonomy is complex: several species groups are recognizable, species from one group replacing each other geographically in the guilds of guenons present in each forest. These species groupings are indicated in the species entries below. Other species occur in suitable habitat over a very large area without obvious racial differentiation. Social organization is varied and is described by species. Adult males give loud species-specific distance calls. Groups of different species may travel together for long periods. Gestation periods have been estimated at around 5 months. Breeding is seasonal where known. Typical birth intervals vary between 1 and 3 years and first births occur between 3–7 years, variation being attributed to species and habitat differences in different cases.

Vervet

Cercopithecus aethiops
Vervet, grivet or Savanna or Green monkey.

Senegal to Somalia and southern Africa. Savanna, woodland edge, never far from water and often on banks of water courses. Semi-terrestrial. Diet: fruit, leaves, flowers, insects, eggs, nestlings, rodents, crops. HBL 18–26in; wt 7lb (female), 10lb (male). Coat: yellowish- to olive-agouti, underparts white, lower limbs gray, face black with white cheek-tufts and browband; eyelids white; scrotum bright blue, penis and perineal patch red. Geographical races have been recognized within their vast range and given specific status according to detail of color and pattern of cheek-tufts, but there is also variation of these characters within one troop. Groups usually include several adult males. Closely related to Diana monkey and Patas monkey.

Redtail monkey

Cercopithecus ascanius
Redtail or Coppertail monkey, Schmidt's guenon.

NE and E Zaire, S Uganda, W Kenya, W Tanzania, SW Rwanda. Mature rain forest and young secondary forest. Arboreal. Diet: insects, fruit, leaves, flowers, buds. HBL 16–19in; wt 7lb (female), 9lb (male). Coat: yellow-brown, speckled, with pale underparts; limbs gray; tail chestnut-red on lower end; face black, bluish around eyes, with white spot on nose and pronounced white cheek fur. Groups often have only one adult male. One of the *C. cephus* species group.

Moustached monkey

Cercopithecus cephus

S Cameroon to N Angola. Rain forest. Arboreal. Diet: fruit, insects, leaves, shoots, crops. HBL 19–22in; wt 6lb (female), 9lb (male). Coat: red-brown agouti with dark gray limbs and back; lower part of tail red; throat and belly white; face black with blue skin around eyes, white mustache bar and white cheek fur. Groups may include only one adult male.

▲ **Vervet monkey troop.** Coat patterns are more distinct in the males, which, particularly in such larger species, may be somewhat larger than females.

Red-eared monkey

Cercopithecus erythrotis

S Nigeria and W Cameroon. Rain forest. Arboreal. Diet: fruit, insects, shoots, leaves, crops. HBL 14–20in; wt 9–11lb (male). Coat: brown-agouti with gray limbs; part of the tail red; face blue around eyes, nose and ear-tips red, cheek fur yellow. One of the *C. cephus* species group.

Red-bellied monkey

Cercopithecus erythrogaster

SW Nigeria. Rain forest. Arboreal. Diet: fruit, insects, leaves, crops. HBL 18in; wt about 13lb (male). Coat: brown-agouti; face black, throat ruff white; belly variable from reddish to gray. A little known species similar to *C. petaurista* and in *C. cephus* species group.

Lesser spot-nosed monkey

Cercopithecus petaurista

Lesser spot-nosed or Lesser white-nosed monkey.

Sierra Leone to Benin. Rain forest. Arboreal. Diet: fruit, insects, shoots, leaves, crops. HBL 14–18in; wt 7lb (female), 7–18lb (male). Coat: greenish-brown agouti; underparts white, lower part of tail red; face black with white spot on nose, prominent white throat ruff and white ear-tufts. One of the *C. cephus* species group.

Owl-faced monkey

Cercopithecus hamlyni

Owl-faced or Hamlyn's monkey.

Zaire to NW Rwanda. Rain and montane forest. Arboreal. Diet: fruit, insects, leaves. HBL 22in; wt ? Coat: olive-agouti with darker extremities; scrotum and perineum bright blue; face black with yellowish diadem and thin white stripe down nose. Lives in small groups with a single male.

L'Hoest's monkey

Cercopithecus l'hoesti

Mt Cameroon and E Zaire to W Uganda, Rwanda. Montane forest. Terrestrial. Diet: fruit, leaves, insects. HBL 18–22in; wt ? Tail hook-shaped at end. Coat: dark gray-agouti with chestnut saddle; underparts dark. The eastern form has a striking white bib, while the western form is less strikingly marked with small bib, light gray cheek fur and whitish mustache markings. Lives in small groups with a single adult male. Includes *C. preussi* (Preuss's monkey).

CONTINUED ▶

Blue monkey
Cercopithecus mitis
Blue, Sykes', Silver, Golden or Samango monkey.

NW Angola to SW Ethiopia and southern Africa. Rain forest and montane bamboo forest. Arboreal. Diet: fruit, flowers, nectar, leaves, shoots, buds, insects; prey includes wood owls and bush babies. HBL 19–26in; wt 9lb (female), 16lb (male). Coat: gray-agouti, with geographic variants often given subspecific rank. The **Blue monkey** (*C. m. stuhlmanni*) has a bluish-gray mantle, black belly and limbs, dark face with pale yellowish diadem; **Silver** (*C. m. doggetti*) and **Golden** (*C. m. kandti*) monkeys are variants with lighter and yellowish mantles, respectively, from W Uganda, Rwanda and E Zaire. **Sykes' monkey** (*C. m. kolbi*) has a chestnut saddle and a pronounced white ruff, from Mt Kenya and Nyandarua. The **samango** of more southern areas is a drab rusty-gray. Live in medium-sized groups of about 20–40 often with only a single adult male. *C. mitis* is replaced in W Africa by the closely similar *C. nictitans*.

Spot-nosed monkey
Cercopithecus nictitans
Spot-nosed or Greater white-nosed monkey or hocheur.

Sierra Leone to NW Zaire. Rain forest. Arboreal. Diet: fruit, leaves, shoots, insects, crops. HBL 17–26in; wt 9lb (female), 14lb (male). Coat: dark olive-agouti; belly, extremities and tail black, face dark gray with white spot on nose. Habits similar to Blue monkey, which replaces it to the west.

Mona monkey
Cercopithecus mona

Senegal to W Uganda. Rain forest. Arboreal. Diet: fruit, leaves, shoots, insects, crops. HBL 18–22in; wt 6–14lb. Coat: back brown-agouti, rump and underparts white; upper face bluish-gray, muzzle pink; hair round face yellowish with dark stripe from face to ear. Lives in fairly large groups which may contain more than one male, or a single adult male. The name comes from the moaning contact call of the females. Similar to Crowned guenon, which replaces it to the south.

Crowned guenon
Cercopithecus pogonias

S Cameroon to Congo basin. Forest. Arboreal. Diet: fruit, leaves, shoots, insects, crops. HBL 18in; wt 7lb (female), 10lb (male). Coat: brown-agouti with black extremities; lower part of tail black; belly and rump yellow; face blue-gray with pink muzzle; prominent black line from face to ear and median black line from forehead forming crest; fur yellow between black lines. Similar in habits to Mona monkey.

De Brazza's monkey
Cercopithecus neglectus

Cameroon to Ethiopia, Kenya to Angola. Swamp forest. Semi-terrestrial. Diet: fruit, leaves, insects. HBL 16–24in; wt 9lb (female), 16lb (male). Coat: gray-agouti with black extremities; tail black; white stripe on thigh and rump white; face black with white muzzle; long white beard and orange diadem; scrotum blue. Lives in small groups, usually a pair with offspring. Freezes when alarmed.

Diana monkey
Cercopithecus diana

Sierra Leone to SW Ghana. Forest. Arboreal. Diet: fruit, leaves, insects. HBL 16–21in; wt about 11lb (male). Coat: gray-agouti and chestnut back; extremities and tail black; white stripe on thigh; rump fur red or cream in different races; face black, surrounded by white ruff and beard. A wide-ranging species of the high canopy, living in medium-sized groups with a single adult male. This species may be allied to Vervet monkey.

Wolf's monkey
Cercopithecus wolfi
Wolf's or Dent's monkey.

A little known species from Zaire, NE Angola, W Uganda, Central African Republic. Arboreal. HBL 18–20in. Includes *C. denti*.

Campbell's monkey
Cercopithecus campbelli

A little known species from Gambia to Ghana. HBL 14–22in; wt 5lb (female), 9lb (male).

Dryas monkey
Cercopithecus dryas

A little known monkey from Zaire.

Genus *Allenopithecus*

A single species separated from *Cercopithecus* because females have periodic (perineal) swelling.

Allen's swamp monkey
Allenopithecus nigroviridis

E Congo and W Zaire. Swamp forest. Habits unknown. Diet: fruit, seeds, insects, fish, shrimps, snails. HBL 16–20in; TL 14–21in; wt? Coat: green-gray agouti with lighter underparts; hair flattened on crown.

Genus *Miopithecus*

A single species, separated from *Cercopithecus* because females have cyclic perineal swelling. Talapoins live in large groups of 70–100 including many adult males. They are sharply seasonal breeders, mating in the long dry season and giving birth 5½ months later. Infants are colored like adults except for the pink face which darkens after about 2 months. The juvenile period is long, with first births occurring at 5 or 6 years.

Talapoin monkey
Miopithecus talapoin

S Cameroon to Angola. Wet and swamp forest, and alongside water courses. Arboreal. Diet: fruit, insects, flowers, crops. HBL 13–14in; TL 14–15in; wt 2lb (female), 3lb (male). Coat: greenish-agouti; underparts and inner sides of limbs pale; scrotum blue; face gray with dark brown cheek stripe.

Genus *Erythrocebus*

A single species separated from *Cercopithecus* because of long limbs and adaptations for running. Patas monkeys live in moderately sized groups, usually with a single adult male. They are seasonal breeders, mating in the wet season and giving birth 5½ months later. Infants are light brown with pink faces which darken by 2 months. The juvenile period is short, with first births occurring at 3 years or even earlier.

Patas monkey
Erythrocebus patas
Patas, Military or Hussar monkey.

Senegal to Ethiopia, Kenya, Tanzania. Terrestrial. Diet: acacia fruit, galls, and leaves; other fruit, insects, crops; gum exudates from trees. HBL 23–29in; TL 24–29in; wt 9–29lb. Coat: shaggy, reddish-brown. Underparts, extremities and rump white; scrotum bright blue; penis red; face black with white mustache; cap brighter red, with black line from face to ear. This species seems closely related to the Vervet monkey.

Macaques

Genus *Macaca*

Heavily built, often partly terrestrial monkeys. The coat is generally dull brownish but the naked skin on face and rump may be bright red; some species have sexual swellings. Tails are up to slightly longer than body length (mostly shorter) or totally absent, depending on species. Males are larger than females, sometimes considerably so. Eclectic diets with fruit as the most common item. Seasonal breeders for the most part, mating in the fall and giving birth in the spring after about 5½ months gestation. Infants have a distinctively colored soft natal coat which is replaced after about 2 months. Macaques live in fairly large groups which may include several adult males. Females generally remain throughout life in their natal group, but males emigrate at adolescence and thereafter live alone, in small groups of males, or in other groups with females for varying periods of time.

Stump-tailed macaque
Macaca arctoides
Stump-tailed or Bear macaque.

E India to S China and Vietnam. Forest, particularly montane. Terrestrial and Arboreal. Diet: fruit, insects, young leaves, crops, small animals. HBL 20–27in; TL 0.5–4in; wt 11lb (female), 17lb (male). Coat: dark brown; face naked, dark red and mottled; rump also naked and dark red. No perineal swelling.

Assamese macaque
Macaca assamensis

N India to Thailand and Vietnam. Forest. terrestrial and arboreal. Diet: fruit, insects, young leaves, crops, small animals. HBL 21–27in; TL 7–15in; wt 13lb (female), 17lb (male). Coat: varying shades of yellowish to dark brown; face and perineum naked, red in adult.

Formosan rock macaque
Macaca cyclopis
Formosan rock or Taiwan macaque.

Taiwan. Terrestrial and arboreal. Diet: fruit, insects, young leaves, crops, small animals. HBL 22in; tail moderately long; wt? Coat: dark brown.

Crab-eating macaque
Macaca fascicularis
Crab-eating or Long-tailed macaque.

Indonesia and Philippines to S Burma. Forest edge, swamp, banks of water courses and coastal forest. Terrestrial and arboreal. Diet: fruit, insects, young leaves, crops, small animals. HBL 15–25in; TL 16–26in; wt 10lb (female), 14lb (male). Coat: varying shades of brown (grayish or yellowish or darker); underside paler; face skin dark gray; prominent frill of gray hair round face. No perineal swelling.

Japanese macaque
Macaca fuscata

Japan. Forest. Terrestrial and arboreal. Diet: fruit, insects, young leaves, crops, small animals. HBL 18–24in; TL 3–5in; wt 18–40lb. Coat: brown to gray; face and rump skin naked, red in adult. No perineal swelling.

Rhesus macaque
Macaca mulatta
Rhesus macaque or Rhesus monkey.

India and Afghanistan to China and Vietnam. Forest, forest edge and outskirts of towns and villages. Terrestrial and arboreal. Diet: fruit, insects, young leaves, crops, small animals. HBL 18–25in; TL 7–12in; wt 12lb (female), 17lb (male). Coat: brown with paler underside; face and rump naked, red in adult. No perineal swelling.

Pig-tailed macaque
Macaca nemestrina

E India to Indonesia. Wet forest. Terrestrial and arboreal. Diet: fruit, insects, young leaves, crops, small animals. HBL 18–24in; TL 5–9in; wt 10lb (female), 18lb (male). Coat: varying shades of brown, with paler underside and darker brown areas around face. Females have large cyclic perineal swelling.

Bonnet macaque
Macaca radiata

S India. Forest, forest edge and outskirts of towns and villages. Terrestrial and arboreal. Diet: fruit, insects, young leaves, crops. HBL 14–24in; TL 19–27in; wt 8lb (female), 14lb (male). Coat: grayish-brown with paler underparts; hair on head grows out in whorl from central crown. No perineal swelling.

Lion-tailed macaque [E]
Macaca silenus

S India. Wet forest. Terrestrial and arboreal. Diet: omnivorous. HBL 18–24in; TL 9–15in; wt 15lb (male). Coat: black with gray around face, in outstanding ruff; tail with slight tuft at tip. Females have cyclic perineal swelling.

Toque macaque
Macaca sinica

Sri Lanka. Wet forest, edges of water-courses, scrub. Terrestrial and arboreal. Diet: fruit, insects, young leaves, crops. HBL 17–21in; TL 18–24in; wt 8lb (female), 12lb (male). Coat: reddish or yellowish-brown with paler underparts; hair on top of head grows out from central crown. No perineal swelling.

Barbary macaque [V]
Macaca sylvanus
Barbary macaque, Barbary ape, Rock ape.

N Algeria and Morocco, introduced to Gibraltar. Mid and high-altitude forest, also scrub and cliffs. Terrestrial and arboreal. Diet: fruit, young leaves, bark, roots, occasionally invertebrates. HBL 20–24in; tail absent; wt 24–33lb. Coat: yellowish-gray to grayish-brown, with paler underparts; face dark flesh colored. Females have dark gray-red cyclic perineal swelling.

Père David's macaque
Macaca thibetana
Père David's or Tibetan stump-tailed macaque.

Tibet to China. Montane forest. Semi-terrestrial. Diet: omnivorous. HBL 24in; TL 2in; wt about 26lb (male). Coat: brown.

Moor macaque
Macaca maura

Sulawesi. Forest. Diet: omnivorous. HBL 26in; tail absent; wt ? Coat: brown or brownish-black; ischial callosities large and pink. Females have cyclic perineal swelling.

Celebes macaque
Macaca nigra

Sulawesi. Forest. Diet: omnivorous. HBL22in; tail absent; wt 22lb (adult male). Coat: black, with prominent pink ischial callosities; face black, prominent ridges down side of nose; hair on head rises to stiff crest. Females have cyclic pink perineal swelling.

Tonkean macaque
Macaca tonkeana

Sulawesi. Forest. Diet. omnivorous. HBL about 24in; tail absent; wt ? Coat: black, lighter brown rump, cheeks; ischial callosities prominent. Females have cyclic pink perineal swelling.

Baboons

Genus *Papio*

The classification of baboons is controversial and several systems have been proposed. Here the Savanna or "Common" baboon is considered to be one species containing three races previously considered separate species. The status of the Guinea baboon is not clear and on behavioral grounds it may be regarded as a western race of the Hamadryas baboon. In fact some latest classifications group all the open country species ("Common", Hamadryas and Guinea baboon) under one species, *Papio hamadryas*. The drill and mandrill are here included in *Papio* (not *Mandrillus* as formerly). Baboons live in large groups. Hamadryas baboons have a hierarchical group structure based on the one-male unit or harem, and this structure may also be present in groups of Guinea baboons, drills and mandrills. The "Common" or Savanna baboons have more informal groups including several adult males. Gestation is about 6 months and breeding is not seasonal. Birth intervals vary around 2 years, depending on the food supply, and first births occur when females are from 4 to 8 years old. Infants have a black natal coat and pink skin for the first 2 months.

Savanna baboon
Papio cynocephalus
Savanna, Chacma, Olive, Yellow or "Common" baboon.

Ethiopia to S Africa, Angola. Savanna woodland and forest edge. Terrestrial. Diet: grass, fruit, seeds, insects, hares and young ungulates, crops. HBL 22–31in; TL 16–24in; wt 26–31lb (female), 46–55lb (male). Coat: gray-agouti, with longer hair over shoulders, especially in adult males; shiny black patch of bare skin present over hips. Females have cyclic perineal swelling. First 3 or 4 tail vertebrae fused in adult giving hook-shaped base to tail. Coat color varies geographically, giving recognizable races which have been previously accorded specific status. The lowland East and Central African form is yellowish (**"Yellow baboon"**), the highland East African form is olive-greenish (**"Olive baboon"**), and the southern African race is dark gray (**Chacma baboon**). The face is naked and black with prominent lateral ridges on the long muzzle especially in adult males. The nose varies geographically, the "Olive" baboon having a pointed nose extending beyond the mouth a little, while the "Yellow" and "Chacma" baboons have retroussé noses.

Hamadryas baboon
Papio hamadryas

Ethiopia, Somalia, Saudi Arabia, S Yemen. Rocky desert and subdesert with some grass and thorn bush. Terrestrial. Diet: grass seeds, roots, bulbs. HBL 30in; TL 24in; wt 22lb (female), 37lb (male). Coat: females and juveniles brown, adult males with silvery-gray cape over shoulders with red naked skin on face and perineum.

Guinea baboon
Papio papio

Senegal to Sierra Leone. Savanna woodland. Terrestrial. Diet: grass, fruit, seeds, insects, small animals, crops. HBL 27in; TL 22in in adult male; wt ? Coat: brown with red naked skin on rump; face brownish red.

Drill [E]
Papio leucophaeus

SE Nigeria and W Cameroon. Rain forest. Terrestrial. Diet: fruit, seed, fungi, roots, insects, small vertebrates. HBL 27in; TL 5in; wt up to 110lb. Coat: dark brown with blue to purple naked rump; face black with white fringe of hair around it. Muzzle long, with pronounced lateral ridges along it. Females much smaller than males.

Mandrill
Papio sphinx

S Cameroon, Gabon, Congo. Rain forest. Terrestrial. Diet: fruit, seeds, fungi, roots, insects, small vertebrates. HBL 31in; TL 3in; wt 25lb (female), 55lb (male). Coat: olive-brown agouti with pale underparts; blue to purple naked rump in adult males, duller in females and juveniles. Face very brightly colored in adult male, with red median stripe on muzzle, ridged side of muzzle blue, beard yellow. Females and juveniles similarly colored but duller. Females much smaller than males.

Genus *Theropithecus*

A single species, the only survivor of an important fossil group. Commonly referred to as a baboon, but very different in vocal and visual communication patterns from *Papio*. Geladas live in large herds within which adult males have harems of several females. Other males live in bachelor groups at the periphery.

Gelada
Theropithecus gelada
Gelada or Gelada baboon.

Ethiopia. Grassland. Terrestrial. Diet: grass, roots, bulbs, seeds, fruit, insects. HBL 20–29in; TL 12–20in; wt 31lb (female), 46lb (male). Coat: brown, fading to cream at end of long hairs; mane and long cape over shoulders; naked area of red skin around base of neck, surrounded by whitish lumps in the female which vary in size with the menstrual cycle. Rump of both sexes also red and naked and rather fat. Muzzle with concave upper line, longitudinal ridges along side of snout. Upper lip can be everted, used in flash display.

Monkeys in Clover

How Rhesus macaques manage to survive in the Himalayas

Most primates inhabit warm tropical and subtropical regions. The macaques, however, have a distribution that includes China, Japan, the Himalayas, North Africa and Gibraltar.

One of the 15 macaques, the Rhesus macaque or Rhesus monkey (*Macaca mulatta*)—well known for the important role it has played in medical research—ranges from Afghanistan through much of India and Indochina to the Yangtze in China, with an isolated population near Pekin.

In northern Pakistan, rhesus live in the mountains up to 13,000ft (4,000m) in temperate forest that is dominated by pines and firs, though deciduous trees such as the maple, horse-chestnut and elm sometimes mingle with the conifers. The climate is highly seasonal. A warm, dry spring gives way to a three-month monsoon season, when about 15in (38cm) of rain falls. Sunshine and clear weather return in the fall, but winter brings freezing temperatures and snow, up to 22ft (6.5m) of it between January and March. How do the monkeys survive in this area?

Himalayan rhesus live in groups of 20–70, each group's home range including 1.2–2.3sq mi (3–6sq km) of rugged terrain. The animals sleep in the trees but spend much of the day on the ground, eating the leaves and roots of herbaceous plants. Clover, in particular, makes up a large proportion of their annual diet. Clover grows only in patches from which the trees have been cleared, either by natural events like avalanches or, nowadays more often, by people. These patches are quite rare, but the monkeys seek them out. They also take advantage of sudden abundances of food items, so that over the year they eat a wide range of foods even though at any one time their diet is narrow. For instance, in spring and early summer they eat young fir tips, wild strawberries and the berries of viburnum, a shrub that grows only in open areas; in summer they relish mushrooms and cicadas; fall brings the cobs of jack-in-the-pulpit and pine seeds buried on the ground in a carpet of dead pine needles. With the onset of winter, snow covers up many potential foods and the monkeys resort to poorer items such as the tough, barely digestible leaves of the evergreen oak, but they still manage to find a few nutritious foods. For example, sweet, sticky sap collects on the needles of some pines trees and they lick it off, probably for its sugar content. Where the snow has melted or blown away, they search for plants with fat roots, which they pull up and eat. In February, the viburnum comes into bloom and they feed heavily on the flowers until the spring thaw. The monkeys are insulated from the cold by a heavy coat that grows in late fall and is shed in the spring. Most animals lose weight, 2 pounds or so, but most winters few die.

In many respects, the social life of the Himalayan rhesus is like that of rhesus in

▲ **On the northern edge** of the species' range in Nepal, this temple-dwelling Rhesus macaque is protected by monks against some of the rigors faced by its cousins in the wild. Patterns of breeding, social life and individual life-history are different from those of Rhesus macaques in the tropics.

◄ **Surviving snowy winters** of northern Japan, the Japanese macaque, another hardy species, is protected by the thick gray coat covering its heavy frame. It is the largest of all macaques.

the tropics. Groups contain 11–70 animals, about half of them adult. Adult females outnumber adult males two to one, though they are born in equal numbers (the death rate is higher in young males than females).

Females spend their lives in the group into which they were born. They can be ranked in a hierarchy, each female's position determined by the number of individuals to whom she cowers in submission. Her daughters all rank immediately below her and above all other females to whom she is dominant. Normally, daughters do not outrank their mothers even when the latter grow old. The rank order of the daughters is determined by their birth order, with the eldest ranking lowest and the youngest achieving highest rank about the time she reaches sexual maturity at about 5 years. Thus changes in the hierarchy usually occur only as daughters approach maturity and outrank their elder sisters. A female's kinship affiliations influence many aspects of her social life. Closely related females move and sit together, groom one another frequently and support each other in fights with other females.

The males, which are somewhat larger than the females and dominant to them, can also be ranked in a hierarchy but their relationships are much less stable. Before reaching sexual maturity at about 7 years, most males leave the group into which they were born and join another, usually neighboring, group. As adults they may transfer again. Genetic studies show that in Pakistan Rhesus monkey social groups are not inbred; this pattern of male emigration helps to prevent it. Some males join a new group by loitering on its periphery, cowering to all the resident males. Others directly challenge the highest ranking male in the group and fight fiercely to establish their position. Serious wounds may be inflicted, and some males possibly die as a result.

Some distinctive features of the Himalayan rhesus' social system reflect the seasonality of their environment compared with the equable climate enjoyed by their southern relatives. In the Himalayas Rhesus monkeys mate only in the fall and most male transfers between groups occur in the preceding three months and, particularly, just before the monsoon. In this dry season, several groups with overlapping home ranges will crowd into valleys which still have running water, and groups often move and feed side by side for hours or days on end. Males seize this chance to transfer. Females give birth every other spring at most, a longer minimal interval between the births of surviving offspring than in Rhesus populations further south (2 years compared to about 8 months). In the Himalayas, the monkeys take about two years longer to mature and probably die younger. Females reach sexual maturity at 5–6 years instead of 3–4 years. In provisioned colonies, animals may live for 28 years, but evidence suggests that in the Himalayas few live beyond the age of 20 years. These aspects of their life history mean that females in the Himalayas usually have few (ie 4 or 5) close living relatives and do not form the large matrilineal groups (20 individuals or more) reported for some rhesus populations. The population's capacity for rapid growth is also reduced.

The survival of Rhesus monkeys in the temperate forests of northern Pakistan depends heavily upon access to plants that grow plentifully only in forest clearings. It is ironic that forest clearance—which if continued will lead ultimately to the disappearance of these monkeys—may in the short run increase their abundance. AFR

Baby Care in Barbary Macaques

Infants as instruments of harmony between adult males

As the one-week-old baby nervously tottered away from its mother, a watchful adult male rushed in and picked it up. Holding it upside down in mid-air, the male "teeth-chattered" and "lip-smacked" at the squealing infant. The noisy exchange drew the attention of three other males (of varying ages), who scrambled to the scene. Huddled together, arms over shoulders, they joined in what seemed to be mutual adulation of the young animal. Grimacing and teeth-chattering, often making purring sounds, they passed the bewildered baby from one of the four to the other. After a few minutes the group interaction ended with one of the adult males grooming the baby nonchalantly as the others resumed their feeding and grooming.

Such infant-directed behavior is quite common in the Barbary macaque. Babies are the focus of a rich repertoire of behaviour among troop members of both sexes and all ages. This male care of babies and "use" of babies in social interactions is notable because males of this species interact more with unweaned youngsters than do males of other Old World monkeys. In common with other macaques, male Barbary macaques protect babies from predators, but unlike other species they also undertake "maternal" chores such as grooming and carrying infants. Among Barbary macaques it is usual to observe close groupings of males (adult, subadult and juveniles) in the presence of babies. Such groupings involve either males taking babies and directly presenting them to other males or males without babies approaching those holding one. The infants themselves are passive participants and males will even use a dead baby or sometimes even an inanimate object during their "group huddling" encounters.

The mating system of these macaques is not very different from that of other macaque species. Because males regularly emigrate from their natal groups at puberty, macaque groups are centered around females who form the permanent core of the social unit. These matrilineal female kin form distinct and cohesive units within the larger group and are usually ranked above or below each other in a hierarchy. Kinship and rank clearly influence the nature of social interactions in the group. Among males normally one is dominant to all the others and is thus able to monopolize copulations with all group females. Babies will then generally be fathered by the leader male. During "group huddling" events he is able to withdraw any infant without any opposition, but males will not choose an infant at random for care-taking or huddling. In fact, two males tend to be involved with each other through their common care-taking relationship with an infant. Whether this relationship is promoted by kinship ties or associations that develop during the breeding season is still not clear. On Gibraltar it appears that those males engaging in care-taking and huddling with particular infants belong to the same matriline as the infant. Males of different matrilines normally use babies in a more random fashion and employ them as "buffers" in a potentially aggressive situation.

The use of babies as buffers is also seen among baboons and other macaques. The form it takes in Barbary macaques follows the same pattern as the non-aggressive huddling encounters. For example, when threatened by another male, a male will pick up a baby and present it to the aggressor. Seemingly appeased, the aggressor will abandon his threatening intentions and join the submissive male in huddling and teeth-chattering over the infant. This very complex behavior is habitual among males when babies are available. When there are no babies nearby, other appeasement behavior, eg socio-sexual mounts (mock copulation between dominant and submissive males), become more frequent.

Proponents of the theory of "agonistic buffering," as this behavior is known, claim

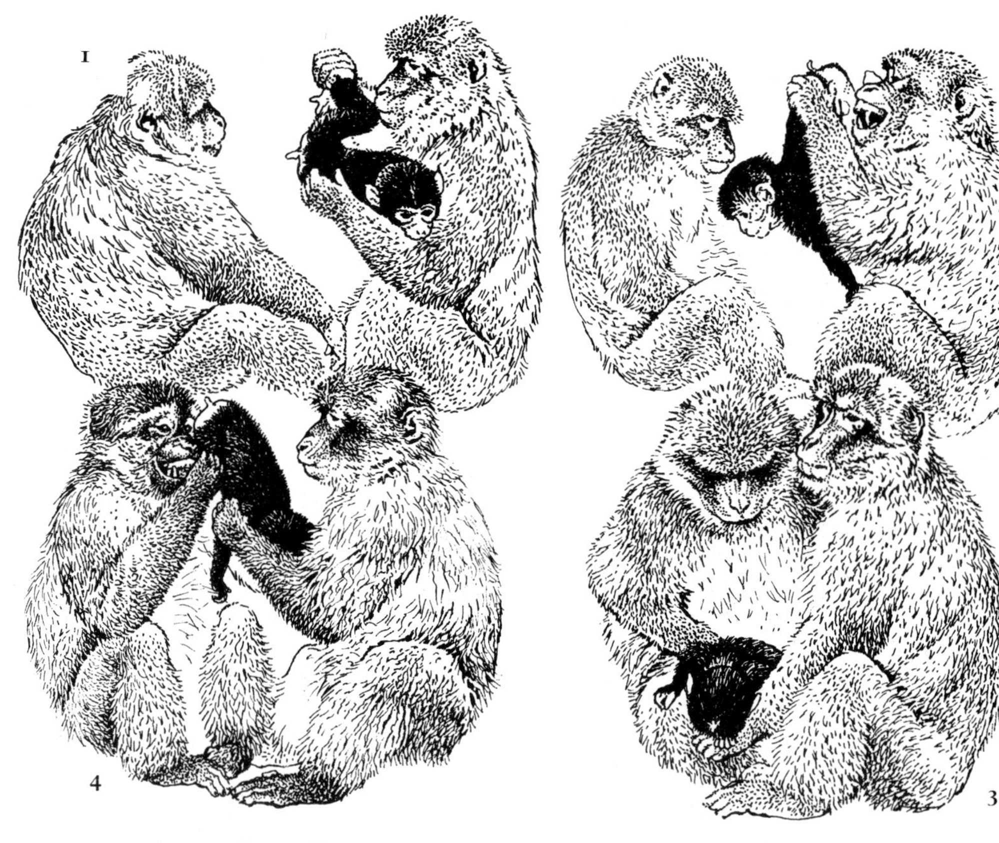

▲ **Adult males** will carry, groom and care for babies, especially if they are related, such behavior often involving two males and an infant. (1) Adult male picks up unweaned infant by hind leg and turns it (2) to sniff genitals, teeth-chattering and lip-smacking as he does so. He holds the infant up to the second male (3) who sniffs, chatters and lip-smacks also. Finally (4) the second male grooms the infant. Mutual grooming between the two adult males may follow.

◀ ▲ **The most famous "Barbary apes"** are on Gibraltar. Barbary macaques have lived on the Rock since at least 1740 when they were imported by the British garrison for game hunting. Numbers have since fluctuated, falling from 130 in 1900 to just four in 1943, when following Mr. Churchill's instructions 24 more were imported from North Africa. Today's population is descended from these individuals and kept at between 30 and 40 in two troops.

that such appeasement interactions allow subordinate animals to gain access to dominant ones and better their chances of "social climbing." But whatever the ultimate consequences may be for the individual's social rank, the behavior is certain to promote a more harmonious social environment. Perhaps the fact that Barbary macaque groups can contain up to 10 adult males as opposed to a few in other macaques is a direct measure of the success of such appeasement rules.

Baby care-taking is an important feature of troop life in any monkey. It has been elaborated in the Barbary macaque not only to promote survival of the infants by giving them constant attention but also to lower the level of tension between animals. JEF

What Does "Whoop-gobble" Mean?

Long-distance communication among mangabeys

The familiar "whoop-gobble" sound heard in the rain forests to the north of the Zaïre River is as stereotyped and distinctive a call as any mammal produces—a low-pitched tonal "whoop," followed by a four- to five-second silence that always ends with a series of loud staccato, "gobbling" pulses—the number and timing of which will identify him individually.

Not until it attains sexual maturity does the male Gray-cheeked mangabey produce for the first time his full-fleged "whoop-gobble" call. To produce his call the male employs both the larynx and additional resonant air sacs. In consequence, the volume of the whoop-gobble easily matches that of most other loud mammal vocalizations, for instance those produced by opera singers, and approaches even that of the sonar pulses of echo-locating bats.

Many forest monkeys have calls analogous to the whoop-gobble: loud, low-pitched, carrying over distances of half a mile or more, and, in some species, given only by adult males. The howls of New World howler monkeys, the whoops of Asian leaf monkeys, the roars of the African Black-and-white colobus, and the booms and pyows of African forest guenons are examples. These calls, though superficially very different in form, are specialized in similar ways to the whoop-gobble and to similar ends. By broadcasting tape-recorded mangabey calls through a speaker and re-recording them some distance away, it has been shown that the whoop-gobble is attenuated less by passage through tropical forest vegetation than is any other mangabey call. One reason is its low pitch, as low-frequency sounds travel well through such a medium. Its stereotyped, distinctive form (other mangabey calls are much more variable than the whoop-gobble) also makes it easy for a monkey's ear to pick out against the inevitable background of tropical bird and insect noises. Over half of mangabey whoop-gobble calls made from 500 yards (0.5km) away are audible to the much less sensitive and attuned human ear. Other monkey species' "loud calls" tend to share the characteristics of low pitch and high stereotypy, presumably for the same reasons.

Most mangabeys give the whoop-gobble call only a few times a day, early in the morning when many birds and other monkeys also call, producing a "dawn chorus." At this time of day the much cooler air below the crowns of the trees tends to focus sounds within the forest canopy. Furthermore, gray-cheeked mangabeys and other species (even those that, unlike these mangabeys, normally live on the ground) tend to give their loud calls from high perches; this is because the further above the sound-absorbent ground low-frequency sounds are

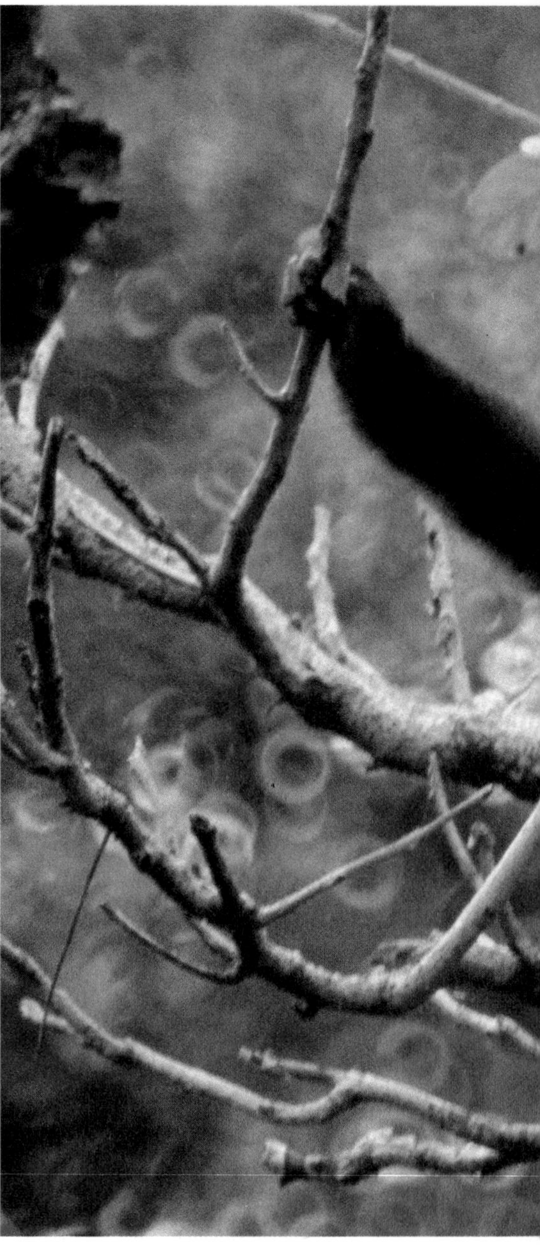

▲ **Air sac inflated** for extra resonance, a male Gray-cheeked mangabey begins his long-distance dawn call. Such "whoop-gobbles" enable neighboring troops to minimize aggressive encounters over food trees.

◄ **Perched high under the forest canopy,** the leading male (1) gives his whoop-gobble. Other troop members continue their activities, foraging for insects or other food (2) or in social activity – (3) female presenting to male. An outsider's whoop-gobble reply from too nearby may elicit a nervous response – (4) male yawning – and the caller will be confronted by the troop leader.

► **Shape of mangabey calls.** Sonograms of the whoop-gobbles of three Gray-cheeked mangabeys (**a–c**). The characteristic whoop-pause-gobble pattern can clearly be seen, as can the differences between individuals.

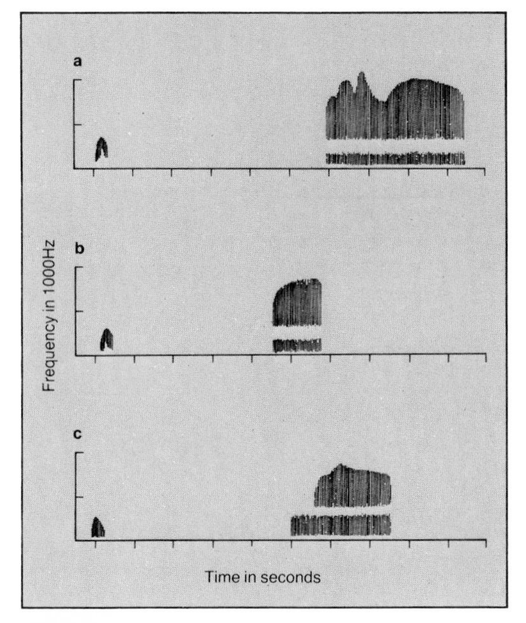

produced, the further they carry.

Gray-cheeked mangabeys are social animals, living in groups of around 15 individuals (range 6–28) including several adult males and a stable core of 5–6 adult females. Unlike some monkeys of more open habitats, they lack a repertoire of more visual signals—bright colors or visual displays would be of limited use in the relatively dark, dense forest canopy where they live. Instead, mangabeys are highly vocal, communicating with each other through a wide variety of grunts, barks, and other acoustic signals as they forage through the trees. The whoop-gobble stands out in the mangabey repertoire by both its audibility and its spontaneity; though sometimes given following a disturbance within the group, it is generally produced without apparent provocation.

If recordings of whoop-gobbles are broadcast near a mangabey group in the field, the mangabeys distinguish between the whoop-gobble calls of males within their group, and can also tell the calls of their "own" males from those of males in other groups. A group of Gray-cheeked mangabeys usually approaches the call of only one of its males, showing no response (except for momentary attention and answering calls) to the whoop-gobbles of its other males. In contrast, the group tends to move away from calls of all outside males. The male whose calls are approached tends to be the most frequent winner of aggressive encounters within the group, and is the most sexually active male; he is also the male most likely to give whoop-gobbles. This individual answers all experimentally broadcast whoop-gobbles (including his own) in kind and may run up to half a mile through the forest to confront the source of a call.

These responses suggest that group members use the call to congregate if they become scattered, and that both males and females can use it to monitor the number and status of adult males in the vicinity. However, the most important listeners are members of neighboring groups: the call is the basic mechanism whereby one group of mangabeys maintains its distance from the next.

The means by which mangabey groups divide the forest among themselves can also be investigated systematically by field playback of tape-recorded whoop-gobbles. A Gray-cheeked mangabey group moves over and feeds in a large home range which, over a year, may cover some 1.6 square miles (4.1sq km) for a 16-member group. The group's response to an intruder is the same—avoidance—whether the intruder's call is from the center of its home range or near its edge. Thus, Gray-cheeked mangabeys do not recognize boundaries or "ownership" of the forest; they are not territorial. However, by advertising their location over long distances and by moving away from the whoop-gobbles of any neighboring group within a few hundred yards, each group maintains a buffer area around itself. As a result, only one group, the group that finds it first, will generally have access to a concentrated food source such as a fruiting tree. Occasionally, due usually to accidents of bad weather or topography, groups encounter one another without hearing each other's calls. In this situation, fighting occurs between the two groups: if the convention of avoidance is broken it is backed up by a threat of real conflict. PMW

A Close-knit Society

Alliances in an Olive baboon troop

The Olive baboons (the name refers to the dark green cast to the grizzled gray-brown coat) of the East African highlands are among the most social of all primates. They live in large troops that occupy home ranges of up to 15.4sq mi (40sq km). The 30 to 150 troop members remain together all the time, as they travel, feed, and sleep as a cohesive unit. It was once thought that it was sexual attraction that kept baboons together, while later observers thought that leadership and authority provided by the adult males formed the basis of group life. It is now known that the enduring relationships formed by females with one another, with young, and with adult males are what lie at the heart of baboon society.

As in many other social mammals, female baboons remain in their natal groups, while males voluntarily leave to join new troops, one by one as they near adulthood. Females and their female offspring often maintain close bonds after the offspring grow up. These associations between maternal kin result in a network of social relationships extending down through three generations and out to include first cousins. Within this female kin group (or matriline, as it is known) each female ranks just below her mother. When a troop of Olive baboons takes a break during the day, relatives will gather around the oldest female in the family to rest and groom. At night relatives usually sleep huddled together, and they will come to one another's aid if a member of the kin group is threatened by another baboon. Females and young from different kin groups within the troop also form close bonds with one another based on years of familiarity. These bonds are developed in play groups, and in nursery groups, associations between females that have infants of similar ages.

A male immigrant who is unfamiliar to members of his new troop therefore has to penetrate a dense network of relatives and friends. He usually begins by cultivating a relationship with an adult female. He will

▲ **Olive baboon troop** feeding on acacia flowers. Although baboons are the most terrestrial and herbivorous Old World monkeys they can and will enter trees in search of food.

◄ **This male appears to be well established** in the troop and to have been accepted as their friend by three, probably related, females.

▼ **A "foot-back" submissive greeting.** In addition to presenting her foot-palm, the female on the left acknowledges inferior social status by a "fear-grin" and the vertically raised tail. Fights between females are rare. Relative status in the female dominance hierarchy that is the basis of Olive baboon society is routinely expressed by such gestures.

follow her about, making friendly faces at her when he catches her eye, lipsmacking and grunting softly, and, if she permits, he will groom her. After many months the male may succeed in establishing a stable bond with a female. If so, their relationship will serve as a kind of "passport," allowing him to gradually extend his ties to her friends and relatives. Although the newly arrived male's ability to compete effectively against other males is an important determinant of his status, if he fails to form such bonds he is unable to stay in the troop for long.

Relationships with females continue to be important for long-term resident males. When female baboons are in heat they mate and interact with many different males, but females spend most of their adult lives either pregnant or nursing. At these times they will not mate. Observation of one large troop of Olive baboons revealed that most of the 35 pregnant or nursing females had a special relationship or "friendship" with just one, two, or three of the 18 adult males (a 1:2 adult sex ratio is typical of this subspecies of *Papio cynocephalus*), and different females tended to have different male friends. Pregnant and nursing females remained near their friends, avoiding all other males, while foraging for grasses, bulbs, roots, leaves, and fruits. While resting during the day or sleeping on the cliffs at night, "friends" often huddled close together. Nearly all of the females' amicable interactions with males, including grooming, were restricted to their "friends."

The main reason why a female develops such "friendships" seems to be the aid the male gives her and the investment that he makes in her offspring. He will sometimes (not invariably) defend the female or her accompanying juvenile offspring from aggression by other troop members. Here the males' strength, powerful canines and size—at 48–80lb (22–35kg) they are twice as heavy as the females—come into play. Male friends also usually help and care for and protect the female's infant. The male will groom and occasionally carry the infant, protecting it from predators as well as troop members. Such bonds between the young and the friend of its mother sometimes persist for years.

Why, for their part, do males form and maintain "friendships" with certain females? In some instances the male friend mated with the female around the time she conceived, in which case he may be the father of her current infant. But close relationships between infants and males who were unlikely to be the infant's father have also been observed. Once a male has demonstrated his willingness and ability to attach himself to a female and her infant, he seems to become more attractive to the female: when a female comes back into heat she prefers to mate with those males who were her friends when she was pregnant and, later, nursing. The male probably befriends the female who is not in heat in order to increase his opportunities for mating when she once again becomes receptive (usually about one year after giving birth).

In the troop discussed here, most females, regardless of age or dominance rank, had just 2 male friends. But older males who had lived in the troop for several years had 5 or 6 female friends, while younger resident males had only 2 or 3, and males who had lived in the troop for less than six months had none. When a male had several female friends they were often from the same matriline. In several troops of Olive baboons some old males mate more often than younger, more dominant males, perhaps because of the older males' greater number of female friends. However, males do compete actively for access to receptive females, and fights are common. Male dominance relationships are less stable than those of linear hierarchies of females in which each kin group has a clear-cut position. Fights between females are rare, status being routinely expressed by submissive gestures.

Natural selection has favored competitive behavior in baboons, but at the same time it has also favored the capacity to develop close and enduring bonds with a few others. The fact that humans share this capacity with other primates suggests that it is a very ancient and fundamental aspect of human nature. BS

A Male-dominated Society

The Hamadryas baboons of Cone Rock, Ethiopia

It is an hour before sunset in the semidesert of the southern Danakil plain in Ethiopia. The long column of Hamadryas baboons, brown females and young interspersed with large, gray-mantled males, crosses the dry river bed and threads its way up through a steep gravel slope. It heads for a cliff where it will pass the night in safety from leopards. Suddenly, a male near the front runs back along the column at full speed. A female separates from the last group in the line and hurries towards him as if she knew that she had lagged too far for his tolerance. Upon reaching her, the male gives her a shaking bite on the back of the neck. Squealing, she follows him closely up to the cliff where his other females are waiting. He then leads his family to a ledge where they settle for grooming.

Hamadryas males herd their females by threats and, in severe cases, by neck bites. Four-fifths of the troop's adult males own harems, which range in size from one to 10 females and average about two. Whereas males of other baboon species consort with only one female at a time and only for hours or days when she is in heat, 70 percent of hamadryas pair bonds last longer than three years and continue uninterrupted through the periods of pregnancy and lactation, when the female is not sexually accessible. They show that primate pair bonds are not necessarily sexually motivated.

The troop which lives at Cone Rock in northeastern Ethiopia numbers several hundred and is organized into groups of four levels—harem families, clans, bands and troop—a structural complexity that is not matched by any other known primate populations apart from man (see RIGHT). Members of the same social unit interact about 10 times more often with each other than they do with "outsiders" belonging to next largest unit.

The reproductive career of a young hamadryas male is successful if he obtains possession of a harem. His difficulty is that all reproductive females in a troop belong to some male who will fight any encroaching rival. Indeed, experiments have demonstrated that males have an inhibition against taking over females from another male in the same band, even when the latter is a *less* powerful fighter. In the band referred to above the subadult follower's standard solution is to court a juvenile female, always in his own clan. Having not reached puberty, she is only moderately defended by her father, and other fully grown males are not interested in her at all. Thus, the subadult may closely precede the juvenile female

► **Hamadryas clan on the move.** The gray mantles of the males (clan leader on RIGHT) stand out among the brown females and young.

► **Intensely possessive towards his harem** BELOW, a male threat-gapes at an opponent during a band fight. The rival turns away, grasping his nearest female around the hip. This possessive frenzy is accompanied by a strong inhibition in the male even to look at another female.

▼ **Smallest social unit is the harem family.** One to three *families*, each with its adult male leader, his females (1–10 in number, but averaging 2) and their young, are escorted by several male followers. The families and their followers form a *clan* of 10–20 baboons, who forage and sleep as a group.

Several clans are united in a *band* of about 70 animals. In the morning at the sleeping cliff, the band's adult males communicate about the direction of the day's foraging march; these interactions may last more than an hour, but eventually band members leave the sleeping site together and later reassemble to drink at noon. The band is also a unit of defense against males of other bands attempting to appropriate some of its females.

The *troop* is an unstable set of bands that use the same sleeping cliff.

The male follower (M) of one family is courting a daughter for his first mate.

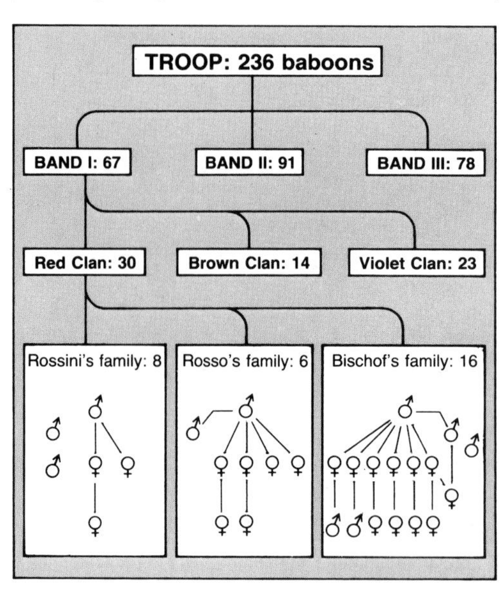

when she follows her mother, and so habituate her to follow *him*. During such maneuvers he must not make her squeal, which might provoke her father to chase him away. With skill, and profiting from the tolerance he enjoys as a clan member, the subadult will extract her from her family in a few months' time. In contrast to this subtle procedure, males in another band always waited till they were young adults and then abducted their juvenile female suddenly and by force: obviously there are diverging band styles of social behavior, even in the same troop. By acquiring a prepubertal female the subadult male avoids the competition of adult males. Once mature she will become his first mate, and the other males will continue to respect the pair bond.

When fully grown, male followers in one band attacked the aging family leaders in their clan and took some of their females by force. Grown males of other clans joined and sometimes obtained females from the disrupted harem. The winning males were younger than the losers. Whereas the clans of the loser's band shared about equally in the lost adult females, only a third of them were lost to other bands. The three defeated old leaders were left with one or no female. Within weeks, they lost weight and their long mantle hair had changed to the hair color of females, in response to what is presumably a very traumatic experience. Only one defeated leader lived on for several years in his clan; he remained most influential in the clan's decisions on travel directions. The other two disappeared: they either left the troop or died.

It appears that the male's reproductive career depends on his association with his clan and band. Clan and band males obtain most of their females from one another and cooperate in their defense against outsiders. In fact, a male remains in his natal clan to the end of his reproductive period. He first becomes one of its followers and finally, with luck, one of its harem leaders. When two mothers of juvenile sons were taken over by males from outside the clan, the sons actually abandoned their mothers and rejoined their fathers' clans, which is an extraordinary preference in a juvenile primate.

Females, however, are frequently transferred to other clans or even bands. Theoretically a juvenile may be expected to join the parent from which he will profit more in his reproductive career. In the more promiscuous baboon species a son does not know who is his father, but may profit from a high rank of his mother. In the hamadryas, he "inherits" females from his father's clan. The hamadryas from Cone Rock are exceptional in imposing something like a patrilineal system on the ancient and widespread matrilineal organization that is typical of related species (see p96).

Nevertheless, females are not merely passive merchandise in hamadryas society. Experiments were necessary to show that a female prefers certain males, and that a rival male heeds her preference: the less a female favors her male, the more likely it is that a rival will overcome his inhibition and rob her. Females, being of far inferior strength and size (some 22lb as opposed to the males' 44lb), must choose more subtle ways to reach their goals, and research naturally first discovered the conspicuous, aggressive stratagems of males. HK

Hybrid Monkeys of the Kibale Forest

Successful mating between Redtail and Blue monkeys

Hybridization between mammal species in the wild is extremely uncommon. Normally, it is avoided or prevented by isolating mechanisms, such as geographical barriers (eg rivers), preference for different habitats in the same region, or differences in the anatomy, physiology, ecology and behavior of species. Most hybridization in the wild occurs where geographical barriers have broken down and two closely related but previously isolated species are able to interact. When there is a breakdown of isolating mechanisms and interbreeding does occur, it is usually disadvantageous, both to the hybrid and the parent species. Most often the offspring are infertile. If not, they may be ill-adapted to their habitat or be unable to find mates.

In Africa the Redtail monkey or Schmidt's guenon (*Cercopithecus ascanius*) and the Blue monkey (*C. mitis*) (see also p88) occur together over much of their geographical ranges without hybridization. However, hybrids have been found in four forests, one of which is the Kibale Forest of western Uganda. This hybridization is of special interest because it occurs between species whose ranges do normally overlap; because the offspring are fertile, and because the female hybrids, at least, seem to have distinct social advantages.

Physically, blues and redtails differ considerably in size and weight (Blue monkeys are larger, often considerably so), color, and markings. Behaviorally, however, they share many similar calls, gestures, and food items, as well as the same type of social system. They both live in harems which consist of a permanent group of adult females with their young, and one temporary adult male. While females usually remain in their natal group for life, males leave before reaching maturity. They become solitary and apparently monitor other social groups, looking for an opportunity to take over a harem. Because adult males are intolerant of one another, there is usually much aggression during takeovers. Male tenure in a harem can last from a few days to several years.

Interactions between these two species in the Kibale Forest are relatively infrequent, despite the fact they are often found together while traveling to and aggregating at food trees. When they do interact, it is usually aggressively, concerning competition over food: the Blue monkey generally supplants the smaller redtail. Behavior that establishes and cements amicable relations, such as grooming and play, is common within each species but not between them. Why then does hybridization occur?

In Kibale, hybridization is confined to the study area of some 2.3sq mi (6sq km) located about half-way up the forest's north-south slope. It is here that the distribution of the Blue monkey population suddenly stops for unknown reasons even though redtails are found throughout the forest. At this southern extreme of the Blue monkey range, their density is only one-seventh of what it is further north, where there are about 14-16 blues per sq mi (35–40/sq km). In fact, there is only one blue social group at Ngogo, but a very high concentration of solitary blue males (at least six, whereas at the same time, in another area further north in the forest, none was observed). Thus, competition by these males for females is extremely intense, resulting not only in a higher rate of male takeovers in this area (six in the Ngogo group and none in four groups further north during the same period) but, apparently, in hybridization.

If a male blue is unable to mate with a female blue, then it seems that the "next best" strategy is to copulate with a closely related species—a female redtail. That this is what has occurred is borne out by the fact that all three known hybrids (one adult

▲ **Confrontation between Redtail monkey troops.** Adult females and juveniles cluster together to face and threaten members of a neighboring troop. A Blue-Redtail hybrid female ABOVE CENTER joins defending the territory of the troop into which she was born. She is larger and more Blue-like than her female relatives. During the tension, individuals often take time out to groom one another RIGHT, perhaps to enhance the strength of the coalition binding the troop. Although threat gestures and vocalizations are intense, physical contact between troops is rare. Harem males are seldom involved in such encounters, and the solitary male Blue monkey TOP, father of the hybrid, adheres to this pattern.

male and two adult females) live in different redtail groups. Presumably they were born and raised in redtail groups as offspring of redtail females and blue males. Both female hybrids have produced offspring and, judging by their appearance, two of the three backcrosses were probably sired by a redtail male, the third by a blue.

In the last case, an adult male Blue monkey joined a redtail group, sharing the "sole" harem male position with a male redtail for more than two and a half years! Although the Blue monkey has successfully mated with the hybrid female in the group (who looks more like a blue), so far the redtail females have tended to avoid him. It is possible that there must be a long period of familiarization before a redtail female would willingly copulate with a particular male blue, especially because she is apparently conditioned to avoid interactions with this species.

The fact that hybrid females are fertile supports the suggestion that hybridization can be a viable "investment" for male blues when competition for females is unusually intense. However, what advantages are there for a redtail female in mating with a blue? For their part, female hybrids appear to be fully integrated into the redtail groups in which they live. They groom and receive grooming and are allowed to handle the small young infants of redtail mothers. In other words, they are not ostracized, but are treated like redtails. Furthermore, their larger size means that they are likely to be dominant to the redtails and therefore have priority of access to food. When assisting the other females in territorial defense against neighboring redtail groups, both size and a blue-like appearance are distinct advantages. Furthermore, hybrids benefit from being able to feed on a wider variety of foods than either of the parental species. They not only eat the foods typical of redtails and blues, but also consume items that neither of them take. And, most importantly, the female hybrids appear to have no trouble mating with either redtail or blue males and producing healthy hybrids.

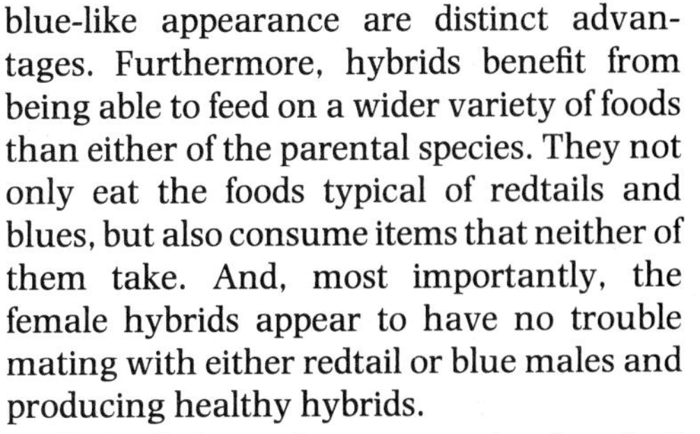

With all these advantages, why do redtail females not hybridize more frequently? The answer may well lie in the reproductive potential of hybrid males. In leaving his natal group at adolescence, the hybrid male also leaves the redtails most familiar with him. In seeking another redtail harem to take over, the hybrid's large size and Blue-monkey appearance will be an enormous disadvantage: redtail females will no doubt treat him with the fear or indifference they would a solitary male blue. As a result, his chances of reproductive success may be considerably lower than those of a redtail male. TTS

COLOBUS AND LEAF MONKEYS

Subfamily: Colobinae
Thirty-seven species in 6 genera.
Family: Cercopithecidae.
Distribution: S and SE Asia; equatorial Africa.

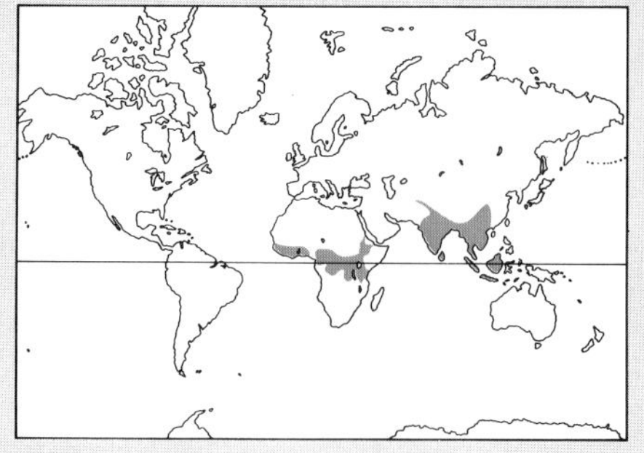

Habitat: chiefly forests; also dry scrub, cultivated areas and urban environments.

Size: from head-body length 17–19.5in (43–49cm), tail length 22.5–25in (57–64cm) and weight 6.4–12.5lb (2.9–5.7kg) in Olive colobus to head-body length 16–31in (41–78cm), tail length 27–42.5in (69–108cm), weight 12–52lb (5.4–23.6kg) in Hanuman langur.

Gestation: 140–220 days depending on species.

Longevity: about 20 years (29 in captivity).

Genus *Nasalis* (2 species): **Proboscis monkey** [V] (*N. larvatus*); **Pig-tailed snub-nosed monkey** [E] (*N. concolor*).

Genus *Pygathrix* (6 species): **Brelich's snub-nosed monkey** (*P. brelichi*); **Tonkin snub-nosed monkey** (*P. avunculus*); **Biet's snub-nosed monkey** (*P. bieti*); **Red-shanked douc** [E] (*P. nemaeus*); **Golden monkey** [R] (*P. roxellana*); **Black-shanked douc** (*P. nigripes*).

Genus *Presbytis* (7 species), including **Mentawai Islands sureli** [I] (*P. potenziani*); **Grizzled sureli** [I] (*P. comata*); **Maroon sureli** (*P. rubicunda*); **Mitered sureli** (*P. melalophos*); **Pale-thighed sureli** (*P. siamensis*); **Banded sureli** (*P. femoralis*).

Genus *Semnopithecus* (13 species), including **Barbe's leaf monkey** (*S. barbei*); **Hanuman langur** or **Common langur** (*S. entellus*); **Dusky leaf monkey** (*S. obscurus*); **Golden leaf monkey** [R] (*S. geei*); **Nilgiri langur** or **Hooded black leaf monkey** [V] (*S. johnii*); **Purple-faced leaf monkey** (*S. vetulus*); **Capped leaf monkey** (*S. pileatus*); **Silvered leaf monkey** (*S. cristatus*).

Black colobus monkeys (4 species of *Colobus*): **Guinea forest black colobus** (*C. polykomos*); **guereza** (*C. guereza*); **Satanic black colobus** [V] (*C. satanas*); **White-epauletted black colobus** (*C. angolensis*).

Red and olive colobus monkeys (5 species of *Procolobus*), including **Olive colobus** [R] (*P. verus*); **Pennant's red colobus** (*P. pennantii*); **Guinea forest red colobus** (*P. badius*).

[E] Endangered. [V] Vulnerable. [R] Rare.
[I] Threatened, but status indeterminate.

Colobus and leaf monkeys—the subfamily Colobinae—exhibit a great diversity of form, and despite an "image" as highly arboreal long-tailed leaf-eating monkeys, less than half regularly subsist on a diet of mature leaves. This diverse assemblage includes the Proboscis monkey, with its haunting appearance and anachronistic display of terrestrial adaptations in an arboreal environment; and the Red-shanked douc monkey, with its ethereal facial skin pigmentation contrasted against the intricate arrangement of four basic coat colors. Equally striking, but less audacious, are the black colobus, notably the guereza. Less dramatic are the Asian leaf monkeys and surelis whose wizened faces have earned one of the genera the name *Presbytis*, meaning "old woman" in Greek.

The Old World monkeys (family Cercopithecidae) are anatomically homogeneous, and few consistent differences distinguish the two subfamilies. The Colobinae are principally distinguished from the subfamily Cercopithecinae by the absence of cheek pouches, and by the presence of large salivary glands and a complex sacculated stomach. The molar teeth have high pointed cusps, and the inside of the upper molars and the outside of the lower molars are less convexly buttressed than they are in the Cercopithecinae. The enamel on the inside of the lower incisors is thicker than in the cercopithecines, and there is a lateral process on the lower second incisor. The sequence of dental eruption differs. Underjet (protrusion of the lower incisors beyond the upper incisors) is common in the colobines, but rare in the cercopithecines.

The majority of living colobines are of slender build compared to the cercopithecines. The two *Nasalis* and six *Pygathrix* species are more thickset and include some of the largest, but perhaps not the heaviest, of monkeys; their fore and hindlimbs are also more equal in length than in other living species. An important feature in the colobines is the trend towards a reduction of the thumb length, least pronounced in the snub-nosed monkeys, the Proboscis monkey and the fossil genus *Mesopithecus*, and most prominent in the colobus, where the thumb is either absent or represented by a small phalangeal tubercle which sometimes bears a vestigial nail. The ischial callosities are separate in females and contiguous in males except in male *Pygathrix* and male *Procolobus*, where the callosities are separated by a strip of furred skin.

The present stronghold of the subfamily is Asia with four genera and 28 species, compared to only two genera and 9 species in Africa. Its fossil representation is strongest in Africa, and two of the earliest fossil genera, *Mesopithecus* and *Dolichopithecus* are European. In Asia, *Pygathrix* and *Nasalis* occupy a zone extending from southern China through eastern Indochina to the Mentawai Islands and Borneo, but curiously are absent from the Malay Peninsula and Sumatra. The rest of the Asian species range from about latitude 35.5°N at the Afghanistan-Pakistan border to the Lesser Sunda Island of Lombok; they are absent from the Philippines and Sulawesi. The living African species are distributed from the Gambia through the Guinea forest belt and the central African forest to Ethiopia, with outlying populations in East Africa and on the islands of Macias Nguema (formerly Fernando Póo) and Zanzibar. Fossil African colobines inhabited northern and southern Africa.

An increasing proclivity for leaves and other plant parts that are less susceptible than fruits to seasonal fluctuations in availability equipped the ancestors of Old World monkeys for survival in open woodland and savanna, which would be inhospitable to frugivores such as hominoids. The cercopithecine diet became more varied, while the colobine diet became more folivorous.

Colobine genera are primarily distinguished on the basis of cranial characters. Variation in newborn (neonatal) coat color is also important and in some genera, dental and visceral anatomy, and external features such as the position of the ischial callosities can be taken into account.

At species level colobines are separated

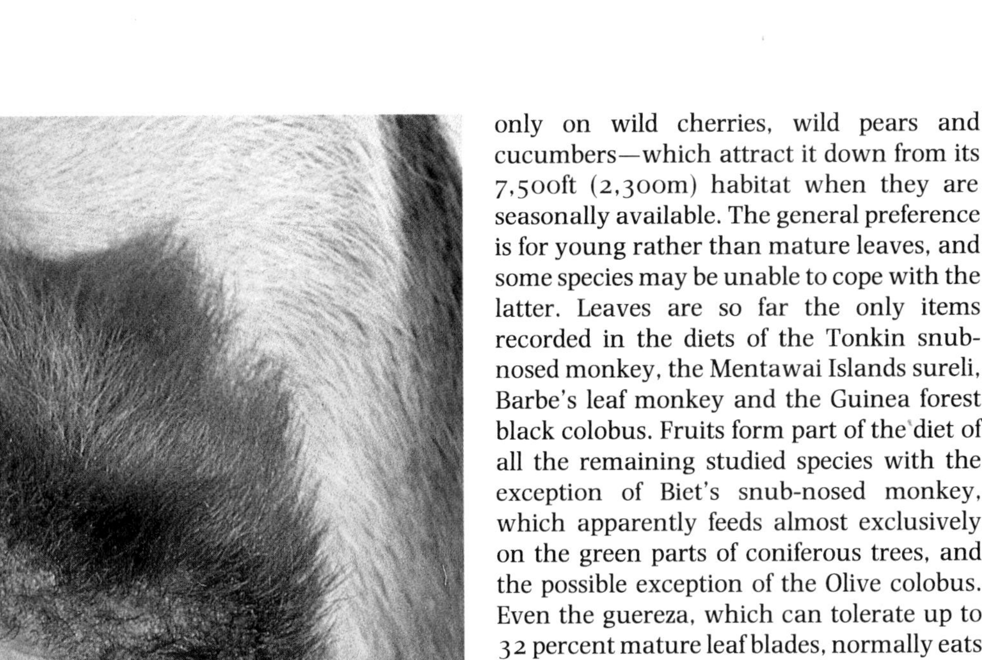

▲ **Tongue-shaped pendulous nose** of the male Proboscis monkey from Borneo contrasts with the snub nose of the female. Males are almost twice the weight of females and are the largest Asian colobines.

◀ **Guinea forest red colobus,** a three-year-old male photographed in the Gambia, westernmost point in the distribution of colobine monkeys. The stump-like reduction of the thumb, a colobine monkey characteristic most marked in colobus species, can just be seen.

chiefly by coat color, but also by length and disposition of the hair (especially on the head, where crests, fringes and whorls may be present) and by vocalization.

The greatest concentrations of colobine species are in Borneo with six species, although not more than five in any one part of the island; and in northeastern Indochina and West and Central Africa, each with three species.

With one exception, all species for which information is available include leaves in their diet. The apparent exception is Brelich's snub-nosed monkey which feeds only on wild cherries, wild pears and cucumbers—which attract it down from its 7,500ft (2,300m) habitat when they are seasonally available. The general preference is for young rather than mature leaves, and some species may be unable to cope with the latter. Leaves are so far the only items recorded in the diets of the Tonkin snub-nosed monkey, the Mentawai Islands sureli, Barbe's leaf monkey and the Guinea forest black colobus. Fruits form part of the diet of all the remaining studied species with the exception of Biet's snub-nosed monkey, which apparently feeds almost exclusively on the green parts of coniferous trees, and the possible exception of the Olive colobus. Even the guereza, which can tolerate up to 32 percent mature leaf blades, normally eats over a third fruits. Most species eat flowers, buds, seeds and shoots. The Hanuman langur, the better studied leaf monkeys, the guereza and red colobus have all been observed to eat soil or termite clay. Golden leaf monkeys specifically eat salty earth or sand, and the Bornean Grizzled sureli churns up the mud at salt springs, and may eat it. Insects occur as a small proportion of the diet of some surelis, of the Hanuman langur and the Hooded black leaf monkey or Nilgiri langur, and probably also of other leaf monkeys, and of Pennant's red colobus. Hanuman langurs and red colobus monkeys eat insect galls and fungi, and African colobines eat lichen and dead wood. Pith occurs in the diets of the Hanuman langur and the Maroon sureli, and roots in those of the Hanuman langur and the Mitered sureli. The latter digs up and eats cultivated sweet potato. The Hanuman langur eats gum and sap, and this species can eat with impunity quantities of the strychnine-containing fruit of *Strychnos nux-vomica* that would kill a Rhesus macaque. It also eats repulsive and evil-smelling latex-bearing plants such as the ak (*Calotropis*) which are avoided by most animals, including insects. Colobines generally get water from dew and the moisture content of their diet, or rainwater held in tree trunk hollows.

Colobines have an unusual stomach whose essential feature is that its sacculated and expanded upper region is separated from its lower acid region. The upper region's neutral medium is necessary for the fermentation of foliage by anaerobic bacteria. The enlargement of the salivary glands indicates their probable role as one of providing a buffer fluid between the two regions of the stomach. The large stomach capacity accommodates the large volumes of relatively unnutritious food and the slow passage essential for fermentation. The stomach contents may constitute more than a quarter of the adult body weight, and as much as half in a semi-weaned infant. The bacterial gastric recycling of urea may be the crucial factor enabling colobines such as the Hanuman langur to survive in arid regions without water sources.

The sacculated stomachs of colobines allow them to digest leaves more efficiently than any other primates. Firstly, the bacteria can break down cellulose (a major component of all leaves) and release energy; primates without such bacteria cannot do this. Secondly, the bacteria can deactivate many toxins and allow the colobine to eat items containing them.

Plant defense compounds are found in all trees, but occur in higher concentration in forests on nutrient-poor soil where it is costly for trees to replace leaves eaten by herbivores. Therefore, in forests growing on good soil, the colobines find the leaves easy to digest and nutritious and they eat mostly leaves of common trees (for example, the red colobus and guereza in Kibale Forest, Uganda). On poor soil, however, the colobines are forced to be more selective; they avoid many common leaves, but eat other plant parts instead, particularly seeds (for example the Satanic black colobus in Cameroon, and the Southeast Asian Pale-thighed and Maroon surelis from Malaya and Borneo respectively).

Female colobines reach sexual maturity at about four years of age, males at four to five years. Copulation is not restricted to a distinct breeding season, but there tends to be a birth peak, timed so that weaning coincides with the greatest seasonal abundance of solid food. Sexual behavior is usually initiated by the female. Receptive female Proboscis monkeys purse the lips of the closed mouth when looking at the male. If he returns her glance she (like the female in *Semnopithecus* species) rapidly shakes her head. If a Hanuman langur female is ignored by the male, she may hit him, pull his fur, or even bite him. The Proboscis male responds by assuming a pout-face and either he approaches the female or she him, presenting her anogenital region. A female Red-shanked douc will characteristically adopt a prone position, and over her shoulder eye the male. He in turn may signal his arousal by intently staring at the female and then turning his gaze to indicate a suitable location where copulation will take place. Soliciting in female *Colobus* is similar but emphasized by tongue smacking. During copulation douc and *Colobus* females remain prone, whereas Proboscis monkey and *Semnopithecus* females adopt the normal cercopithecid quadrupedal stance. The Proboscis female continues to head-shake, and both partners show the mating pout-face.

At birth infants are about 8in (20cm) in head-body length and weigh about 0.9lb (0.4kg). The eyes are open and the infant can cling to its mother strongly enough to support its own weight, although the Olive colobus infant may be carried in its mother's mouth. Body hair is present, but is shorter, more downy, and usually of a different color than in adulthood. There is usually less pigment in the skin and ischial callosities than in the adult, but in the facial skin of the Proboscis monkey and the Red-shanked douc, the opposite is the case. Births are single or, rarely, twin. Parental care in all species so far studied, except for the Pale-thighed sureli and the red colobus, involves toleration by the mother of her offspring being carried off by

► **Hanuman langur mother and young.** Adult size is attained at about five years of age. The single newborn infant may be cared for by temporary female "baby-sitters" and the mother may even nurse infants other than her own.

▼ **The Golden leaf monkey** of Assam and Bhutan. Orange coloration, more marked on the underside, and in adult males and in winter, the crown tufts and the all-black face are characteristic.

other females. Soon after birth, the infant is usually handled and carried by several females to as far as 75ft (25m) from its mother. A mother may even suckle the infant of another female and her own simultaneously, and one has even been observed carrying three infants. Often, a "babysitting" female will abandon her charge and move off, leaving the screaming infant to find its way back to its mother. In *Semnopithecus* active rejection of the young has been observed at the early age of about five weeks. One suggestion is that these behavior patterns accelerate the infant's independence, enhancing its chances of survival during the high infant mortality which often accompanies the violent replacement of the adult male in the one-male troops which occur in some populations. They may also allow the mother to concentrate on time-consuming foraging. Transition from neonatal to adult coat occurs at 5–10 months of age; adult size is attained at about the fifth year.

Compared with macaques, the social relationships of the colobines are typified by a generally lower level of aggressive, sociosexual, vocal and even gestural interactions. Their "grave and serious" demeanor, which prompted the generic name *Semnopithecus* (Greek for "sacred ape"), is probably related partly to their predominantly arboreal habitat which demands less troop coordination for the evasion of predators; and perhaps primarily to their foraging behavior which, owing to the more homogeneous distribution of their food, is less nomadic, requires less acquired experience of choice food sources, and entails prolonged periods of sedentary feeding. Entry into, or movement within, a feeding tree is characterized by meticulous care to avoid close encounters with companions already stationed there. Once a feeding position is attained, its very nature, facing towards the periphery of the tree where the bulk of the food is located, enables the colobine to feed for long periods without interacting with its neighbors. Although commonly there is a feeding peak in the morning and another in the late afternoon, the low nutritional content of their diet usually necessitates intermittent feeding throughout the day, thus minimizing opportunity for complex social behavior.

Group sizes range from solitary animals, usually males, to a group of over 120 langurs (possibly a temporary aggregation of troops seeking water). There are reported groups of a hundred or more Golden snub-nosed monkeys, but such group sizes exceed the maximum recorded for the Satanic black and Pennant's red colobus by only about

◀ **High in the Himalayas,** this large, white-fronted Hanuman langur inhabits the Helenbu Valley 60mi north of Katmandu, Nepal. Langurs and leaf monkeys are the largest, most widespread and diverse group of Asian colobines.

▼ **A troop of guerezas** or White-mantled black colobus feeding in the trees on the shore of Lake Naivasha, Kenya. The U-shaped mantle of long white hairs over glossy black, and the white tail tip distinguish the species.

20. Groups of 60 or so have been reported for the Proboscis monkey, the Red-shanked douc and the Guinea forest red colobus; and troops of up to 40 for most genera. However, with the exception of Pennant's red colobus, where the typical number is probably 50, the average troop size is lower, ranging from 3.4 in the Mentawai Islands sureli, which is unique in apparently being exclusively monogamous, to 37 in one Hanuman langur population. In the Pig-tailed snub-nosed monkey and the White-epauletted black colobus it is about five; in most of the surelis, the Hooded black, Purple-faced, and Capped leaf monkeys and the guereza it ranges from 6 to 9; and in most of the remaining species from 10 to 18. The representation of adult males in a bisexual troop is roughly proportional to its size, and is equalled or exceeded in number by adult females. The Pig-tailed snub-nosed monkey commonly practices monogamy (possibly owing to human predation); the Black-shanked douc in eastern Cambodia is invariably encountered in (presumably bisexual) pairs; and the Bornean Grizzled sureli often occurs in "family parties" of three. The presence of all-male troops has been confirmed only in the Grizzled and Banded surelis, the Hanuman langur, and the Hooded black and the Golden leaf monkeys.

Home range size in most of the species for which it has been estimated is about 74 acres (30ha). It has been estimated for the Capped leaf monkey at 158 acres (64ha), and for the Proboscis monkey at about 320 acres (130ha) which is also the upper limit for Pennant's red colobus. In the Hooded black leaf monkey it ranges from 15 to 642 acres (6–260ha), and the Hanuman langur from 12 to 3,200 acres (5–1,300 ha). On the basis of defense and exclusive use of at least the major part of their home range, the Grizzled and the Pale-thighed surelis, the Purple-faced, the Silvered and the Dusky leaf monkey, and some Hanuman langur populations are considered territorial. Ceylonese langur troops often temporarily desert their home range in order to attack an adjacent troop; while the adult male Purple-faced leaf monkey has such a fastidious sense of territory that it has been seen to chastise fellow troop members for transgressions into other territories. Other Hanuman langur populations, and other species such as the Mentawai Islands sureli, the Hooded black and the Capped leaf monkey, have exclusive core areas which include important sleeping and feeding trees, and which occupy 20 to 50 percent, or in a red colobus troop studied in the Gambia, 83 percent, of the home range. Within its home range the guereza has a preferred area from which other troops are readily chased, but not permanently excluded. In contrast, three Pennant's red colobus troops were found to have very extensive, if not complete, home range overlap. Relations between these three troops were usually aggressive, involved only the adult and the subadult males, and no matter where an encounter occurred within their home ranges, one dominant troop usually supplanted the other, although there were indications that the outcome of such an encounter might also depend on which males were involved. Other red colobus troops entered the area very infrequently and were usually chased out immediately.

Surelis, langurs and black colobus in

particular are characterized by their loud calls which are generally most intense and most contagious at dawn, but may also be heard during the day, especially during preparation for troop movement, finalization of night-time sleeping positions and, in some species, during the night. These calls are believed to promote troop cohesion and to enable prediction of the following day's movement by adjacent troops so that inadvertent encounters can be avoided. During an intertroop encounter or sometimes on detection of a predator, these calls may be preceded or accompanied by a dramatic leaping display in which the protagonist plunges onto branches and then ascends and repeats the performance, producing both a visual and an auditory effect from the swaying and the cracking of branches.

Population densities are very variable, both within and between species. In the surelis, they range from 7 to 125 animals/sq mi (3–48/sq km); in the Pig-tailed snub-nosed monkey and the leaf monkeys 20–570 animals/sq mi (8–220/sq km); in the Hanuman langur 7–2,340 animals/sq mi (3–904/sq km); and in the colobus 70–2,280 animals/sq mi (30–880/sq km). In the Hanuman langur lower densities are most typical of populations living in open grassland or agricultural fields where troops have large home ranges. Intermediate densities appear typical of populations living in close association with towns and villages, and higher densities typical of forest-dwelling populations where the troops have small, sometimes overlapping, home ranges.

In 1819, a single ship's crew visiting Da Nang, Vietnam, killed more than 100 douc between 5am and breakfast-time. Apparently unmolested by the local people, the monkeys lacked respect for firearms and were actually drawn to their deaths by the cries of their wounded companions. Depredations by visiting Europeans were such that by 1831 doucs had learnt to flee from gunfire. In 1974, after depletion of habitat and the ravages of war, only 30–40 doucs were found during 10 weeks in the same locality.

The Guinea forest black colobus was seriously threatened by the European fur trade at the end of the 19th century, and the guereza is still threatened by a tourist demand for rugs and wall-hangings made from their pelts. Other species hunted for their beautiful coats are the Hooded black leaf monkey and the Golden snub-nosed monkey. The coat of the latter was said to protect the wearer from rheumatism, but fortunately only officials and members of the Chinese Imperial family were allowed to wear it. Both species are now protected, and the former is said to have increased in numbers since 1960. Most species are hunted by local people for their flesh which in many species is reputed to have medicinal value, as are the besoar stones (intestinal concretions resembling those found in ruminants) reported in all the Bornean colobines except for the Banded sureli and the Silvered leaf monkey. Nevertheless, even with the wider availability of firearms, the most insidious threat to the future survival of the colobines is the relentless destruction of habitat for timber and for agriculture. It is essential that at least part of the remaining

▲ **White "spectacles"** and bluish face with pinkish lips and chin are characteristic of the lively Dusky leaf monkey from northeast India to the Malay Peninsula and neighboring islands.

► **Leaping a canal** in Gujarat state, Western India, this Hanuman langur is at home in town and temple, revered by Hindu and Buddhist alike.

China's Endangered Monkeys

One of the most endangered and least-known of China's primates is the Golden monkey (*Pygathrix roxellana*) first described in the late 19th century. Mainland China has three *Pygathrix* species. The most endangered is the Black snub-nosed golden monkey or Biet's snub-nosed monkey (*P. bieti*), inhabiting Yunnan Province. Its estimated population is 200 animals. The White-shoulder-haired snub-nosed golden monkey or Brelich's snub-nosed monkey (*P. brelichi*) inhabits the Fan-Jin Mountains in Kweichow Province and totals 500 animals. The Golden monkey is the most numerous and widely distributed species and inhabits Gansu, Hubei (notably Shen Nong Jia), Shaanxi and Szechwan Provinces and totals 3,700–5,700 animals.

Golden monkeys inhabit mountain forests in some of the largest troops known for arboreal primates. Troops of over 600 animals have been reported. In ecologically disturbed areas troops may number 30–100 animals. Larger troops are organized around polygynous subgroups of 1 adult male, 5 adult females and their offspring. There are also peripheral and solitary males, but within the troop adult females outnumber adult males. Males defend the troop against predators (chiefly Yellow-throated martens).

Golden monkeys are basically leaf-eaters, supplementing their diet with fruit, pine-cone seeds, bark, insects, birds, and birds' eggs. Because they lack cheek pouches and eat leaves, they feed often and in large quantities. In zoos they are fed 2–3lb (850–1,380g) daily. There are difficulties in providing a balanced diet, and they fare poorly in captivity. No Golden monkeys are kept in captivity outside China.

Humans have valued the species' decorative shoulder and back hair (up to 14in/10cm long) for making coats for over a thousand years, and herbal medicines are made from the meat and bones. But the major threat to the Golden monkey's survival was vast destruction to their limited habitat. The Chinese government has taken steps to preserve this rare and beautiful primate. Preservation areas have been established, although there is no reserve for the most endangered species, Biet's snub-nosed monkey; hunting was banned in 1975 and forest destruction has been stopped. Anyone breaking the law is fined and may be imprisoned. As a result, births are increasing in preservation areas, allowing guarded optimism, but long-term studies are urgently needed. FEP

habitat should be protected by the establishment of secure reserves; the areas at present most critically requiring protection are the Mentawai Islands and north Vietnam.

There is little contact between colobines and man, excepting the Hanuman langur which in some areas obtains 90 percent of its diet from agricultural crops, and is considered sacred by Hindus and, like all animals, is unmolested by Buddhists. Its sacred status stems from its identification with the monkey-god Hanuman who played the major role in assisting the incarnate god Vishnu in the search for the recovery of his wife, who had been kidnapped by Ravana of Sri Lanka. While in Sri Lanka, Hanuman stole the mango, previously unknown in India. For this theft he was condemned to be burnt, and while extinguishing the fire he scorched his face and paws which have remained black ever since. Many Hindus regularly feed langurs, generally every Tuesday, which is Hanuman's traditional day. The frustration experienced by those whose crops are decimated or have their shops pilfered was well expressed by the town which in desperation dispatched a truckload of langurs to a destination several stations down the railway line. DB-J

THE 37 SPECIES OF COLOBUS AND LEAF MONKEYS

Abbreviations: HBL = head-and-body length. TL = tail length. wt = weight.
[E] Endangered. [V] Vulnerable. [R] Rare. [I] Threatened, but status indeterminate.

Genus *Nasalis*

Thickset build; macaque-like limb proportions, skull shape, and coat color: nose prominent.

Proboscis monkey [V]

Nasalis larvatus

Borneo, except C Sarawak. Tidal mangrove nipapalm-mangrove and (mainly riverine), lowland rain forest. Swims competently. HBL 21–30in; TL 20–29in; wt 18–52lb; Adult male about twice weight of adult female. Coat: crown reddish-orange with frontal whorl and narrow nape extension flanked by paler cheek and chest ruff; rest of coat orange-white or pale orange, richer on lower chest, variably suffused with gray flecked with black and reddish on shoulders and back; triangular rump patch adjoining tail; penis reddish-pink; scrotum black. Elongated nose in adult male is tongue-shaped and pendulous. Newborn have vivid blue facial skin.

Pig-tailed snub-nosed monkey [E]

Nasalis concolor

Pig-tailed snub-nosed monkey or langur, or Pagai Island langur or simakobu.

Mentawai Islands. Rain forest; mangrove forest. Also known as *Simias concolor*. HBL 18–22in; TL 4–7in; wt 16lb. Coat: blackish-brown, pale-speckled on nape, shoulder and upper back; white penal tuft. Face skin black bordered with whitish hairs. Tail naked except for a few hairs at tip. 1 in 4 individuals are cream-buff, washed with brown.

Genus *Pygathrix*

Large with arms only slightly shorter than legs; face short and broad; shelf-like brow ridge; region between eyes broad; nasal bones reduced or absent; nasal passages broad and deep; small flap on upper rim of each nostril.

Golden snub-nosed monkey [R]

Pygathrix (*Rhinopithecus*) *roxellana*

Golden or Orange or Roxellane's monkey or snub-nosed monkey, or Moupin langur.

In Chinese provinces of Hubei, Shaanxi, Gansu and Szechwan. High evergreen subtropical and coniferous forest and bamboo jungle which is snow clad for more than half the year; migrates vertically biannually. HBL 26–30in; TL 22–28in; wt ? Coat: upperside and tail dark brown or blackish, darkest on nape and longitudinal ridge on crown; underside, tail tip and long (4in) hairs scattered over shoulders whitish-orange; legs, chest band and face border richer orange; orange suffusion throughout increases with age in adult male. Muzzle white; areas above eyes and round nose pale blue; colors in young paler.

Biet's snub-nosed monkey

Pygathrix (*Rhinopithecus*) *bieti*

Biet's or Black snub-nosed monkey.

Yun-ling mountain range (26.5–31°N), Yunnan and Tibet. High coniferous forest (11,000–13,200ft) with frost for 280 days per annum. HBL about 29–33in; TL about 20–28in; wt ? Coat: blackish-gray above as in *P. roxellana*, but including paws (paler), brow, inside of limbs below elbow across chest and from hip across abdomen; long yellowish-gray hairs with black-brown tips scattered on shoulders; longitudinal crest, paws, 2/3 of tail and some hairs on upper lip blackish; rest of coat whitish. Placed by some in *P. roxellana*.

Brelich's snub-nosed monkey

Pygathrix (*Rhinopithecus*) *brelichi*

Brelich's or White-shoulder-haired snub-nosed monkey.

Fan-jin Mountains, Kweichow Province, China. Evergreen subtropical forest. HBL about 29in; TL about 38in; wt ? Coat: upperparts grayish-brown, pale gray on thigh; tail, paws, forearm and outside of shank more blackish; tail tip and blaze between shoulders yellowish-white; nape and vertex whitish-brown suffused with blackish, especially at front and sides; underside pale yellowish-gray; ridge of brow hair round face; midline parting from brow to crestless vertex; tail hairs sometimes long with midline parting.

Tonkin snub-nosed monkey

Pygathrix (*Rhinopithecus*) *avunculus*

Tonkin or Dollman's snub-nosed monkey.

Bac Can and Yen Bai, N Vietnam. Bamboo jungle. HBL 20–24in; TL 26–36in; wt ? Coat: upperside blackish; brown between shoulders; occasionally sprinkled with white; nape and rear of crown brown or yellowish-brown with narrow blackish-brown border at front and sides; paws blackish-brown; yellowish-white to orange underside, throat and chest band, almost encircling ankle and hip; tail blackish-brown with whitish-yellow or orange-gray tip; tail hairs without parting, or long with midline parting or helical parting; crown hairs flat. Penis black; scrotum white-haired.

Red-shanked douc monkey [E]

Pygathrix nemaeus

Red-shanked douc or Cochin China monkey.

C Vietnam and E C Laos. Tropical rain and monsoon forest. HBL 21–25in; TL 22–26in; wt? Coat: white lips, cheeks and throat, inside of thigh, perineum, tail and small triangular rump patch; white areas surrounded by black often with intervening deep orange band, most conspicuously between throat and chest; crestless crown, upper arms and trunk between black areas black-speckled gray, forearm white, rest of shank deep orange-red. Penis reddish-pink; scrotum white; skin of muzzle white, ears, nose and rest of face orange.

Black-shanked douc monkey

Pygathrix nigripes

Black-shanked or Black-footed douc monkey.

S Vietnam, S Laos and E Cambodia. Tropical rain forest; gallery and monsoon forest. HBL 22–28in; TL 26–30in; wt ? Distinguished from *P. nemaeus* by anatomy of palate and black-speckled gray forearm and blackish shank. Penis red; scrotum and inside of thigh blue. Facial skin blue, with reddish-yellow tinge on muzzle. Placed by some in *P. nemaeus*.

Surelis

Genus *Presbytis*

Forearm relatively long; brow ridges usually poorly developed or absent; bridge of nose convex; muzzle short; 5th cusp on lower 3rd molar usually reduced or absent; cusp development on upper 3rd molar variable; projection on inner face of relatively broad, underjetted lower incisors; longitudinal crest; coat of newborn whitish.

Mentawai Islands sureli [I]

Presbytis potenziani

Mentawai Islands or Red-bellied sureli.

Mentawai Islands. Rain, mangrove forest. HBL 17–23in; TL 20–25in; wt 12–16lb. Coat: small ridge-like crest; upperside and tail blackish; pubic region yellowish-white; brow band, cheeks, chin, throat, upper chest and sometimes tail tip whitish; rest of underside and sometimes collar reddish-orange, brown or occasionally whitish-orange.

Grizzled sureli

Presbytis comata

Grizzled or Gray or Sunda Island sureli.

W Java, N and E Borneo and N Sumatra. Tropical rain forest. Also named (wrongly) *P. aygula*. HBL 17–24in; TL 22–33in; wt 12–18lb. Coat: paws blackish; crown blackish or brownish; in Borneo adult sexual dimorphism in crest shape and extent of white on brow; rest of upperside pale gray speckled with blackish or brownish; underside whitish. Includes *P. thomasi*, *P. hosei*.

White-fronted sureli

Presbytis frontata

E, SE and C Borneo. Tropical rain forest. HBL 16–24in; TL 25–31in; wt 12–14lb. Coat: paws, cheeks and brow blackish; forearm, shank and sometimes tail base and crest blackish-brown; trunk pale grayish-brown, yellowish below, tail yellowish, speckled with dark gray. Tall, compressed crest raked forward, with 1–2 whorls at base, flanked by laterally directed fringe.

Banded sureli

Presbytis femoralis

Malay Peninsula, C Sumatra, Batu Islands, NW Borneo. Rain forest, swamp and mangrove swamp. HBL 17–24in; TL 24–33in; wt 13–18lb. Coat: dark brown or blackish with variation in spread of whitish underside. 0–2 frontal whorls. Placed by some in *P. melalophos*.

Pale-thighed sureli

Presbytis siamensis

Riau Archipelago, S Malay Peninsula, E C Sumatra, Great Natuna Island. Tropical rain and swamp forest. HBL 16–24in; TL 23–33in; wt 11–15lb. Coat: limb extremities and brow blackish; outside of thigh grayish/whitish; rest of upperparts and tail pale grayish or blackish brown; underparts whitish; horizontal fringe radiates from 0–2 whorls at front end of crest. Placed by some in *P. melalophos*.

Mitered sureli

Presbytis melalophos

Mitered or Black-crested sureli or simpai.

SW Sumatra. Tropical rain and swamp forest, village centers; HBL 16–23in; TL 24–32in; wt 13–16.5lb. Coat: back (occasionally) broad midline band only) brownish-red to pale orange, or gray, variably suffused with blackish or gray; brow reddish to whitish, often delineated by blackish, brown or gray crest and brow hairs which may extend to ear; frontal whorl; underside whitish or whitish suffused with yellow or orange, especially chest and limbs.

Maroon sureli

Presbytis rubicunda
Maroon or Red sureli.

Karimata Island and Borneo, except C Sarawak and lowland NW Borneo. Rain forest. HBL 18–22in; TL 25–31in; wt 11–17lb. Coat: blackish-red or reddish-orange, paler underside, more blackish or brownish on tail tip and paws; horizontal fringe radiates from 0–2 frontal whorls; 1–2 nape whorls sometimes present.

Langurs and leaf monkeys

Genus *Semnopithecus*

Included in *Presbytis* by some. Brow ridge shelf-like and coat of newborn blackish in Malabar and Hanuman langurs (subgenus *Semnopithecus*). Brow ridges resemble raised eyebrows, newborn's coat orange or whitish suffused with gray, brown or black in all other species (subgenus *Trachypithecus*).

Malabar langur

Semnopithecus hypoleucos

SW India between W Ghats and coast to 14°N. Evergreen forest, cultivated woodland and gardens. HBL 24–27in; TL 33–36in; wt 18–25lb. Coat: paws and forearm blackish; leg blackish or grayish-brown; tail blackish or dark gray at base, tip blackish, yellowish-gray or white; midline of back dark brown or dark gray; flanks and rear of thigh pale yellowish-gray; crown orange-gray or yellowish-white; throat and underside orange-white. Frontal whorl.

Hanuman langur

Semnopithecus entellus
Hanuman or Common or Gray langur.

S Himalayas from Afghanistan border to Tibet between Sikkim and Bhutan; India NW of range of Malabar langur to Aravalli Hills and Kathiawar, and NE to Khulna province, Bangladesh; N, E and SE Sri Lanka. Reported in W Assam. Forest, scrub, cultivated fields, village and town centers (0–13,400ft). HBL 16–31in; TL 27–42in; wt 12–52lb. Coat: upperparts gray or pale grayish-brown, often tinged with yellowish; in Bengal gray almost replaced by pale orange; crown, underparts and tail tip whitish or yellowish-white; paws often, and forearms occasionally blackish or brownish; in Sri Lanka and SE India crown usually crested. Frontal whorl.

Purple-faced leaf monkey

Semnopithecus vetulus
Purple-faced leaf monkey or wanderoo.

SW, C and N Sri Lanka. Forest, swamp, rocky, treeless, coastal slopes, parkland. Formerly known as *S. senex*. HBL 18–27in; TL 24–36in; wt 9–22lb. Coat: brown, darkest at limb extremities and sometimes with yellow to brown tail tip and crown, or blackish with pale brown crown and yellowish tail tip; white to yellow throat and sideways-directed whiskers; rump patch sometimes whitish or yellowish; tail base, thighs and back sometimes gray-speckled.

Hooded black leaf monkey [V]

Semnopithecus johnii
Hooded or Leonine or Gray-headed black leaf monkey, or Nilgiri langur.

W Ghats of India and Cat Ba Island, N Vietnam. Evergreen and riverine forest; deciduous woodland; on Cat Ba stunted tree-clad limestone hills. HBL 20–29in; TL 27–38in; wt 21–30lb. Coat: yellowish vertex grades (further in Cat Ba) through brown to glossy brown-tinged black of rest of body; gray-speckled short-haired rump coloration sometimes extends to thigh and tail.

White-headed black leaf monkey

Semnopithecus leucocephalus

SW Kwangsi Province, China. Tropical monsoon forest on limestone hills. HBL 18–24in; TL 30–35in; wt 17–21lb. Coat: glossy black; head and shoulders white; paws and distal part of tail whitish; pointed coronal crest. Placed in *S. francoisi* by some.

White-rumped black leaf monkey

Semnopithecus delacouri
White-rumped or Pied or Delacour's black leaf monkey.

N Vietnam. Limestone mountains. HBL 22–33in; TL 32–34in; wt ? Coat: glossy black with white from end of mouth to nape whorl; sharply demarcated white area on hindpart of back and outside of thigh; tail base hairs long; pointed coronal crest. Placed by some in *S. francoisi*.

White-sideburned black leaf monkey

Semnopithecus francoisi
White-sideburned or François's black leaf monkey.

NE Vietnam, Kwangsi and Kweichow. Tall riverside crags, tropical monsoon forest on limestone mountains. HBL 20–26in; TL 32–35in; wt 13lb. Coat: glossy black with white from end of mouth to ears; pointed coronal crest; 2 nape whorls.

Ebony leaf monkey

Semnopithecus auratus
Ebony or Moor or Negro leaf monkey.

NW Vietnam, Java, Bali and Lombok. Forest; plantations. HBL 18–29in; TL 24–32in; wt ? Coat: glossy black, tinged with brown, especially on underside and cheeks; whitish sometimes on paws. Treated by some as same species as *S. cristatus*.

Silvered leaf monkey

Semnopithecus cristatus

Sumatra, Riau-Lingga Archipelago, Bangka, Belitung, Borneo, Serasan, W coastal W Malaysia, S C Thailand, Cambodia and S Vietnam. Forest swamp, bamboo, scrub, plantations, parkland, village centers. HBL 16–24in; TL 23–33in; wt 11–19lb. Coat: brown, brownish-gray or blackish-brown, darker on paws, tail and brow; color masked by grayish or yellowish hair tips; groin and underside of tail base yellowish; coronal crest variably developed.

Barbe's leaf monkey

Semnopithecus barbei

S China, N Indochina into Burma. Forest. HBL 17–24in; TL 24–35in; wt 10–19lb. Coat: gray to blackish-brown; paws and brow black or blackish-brown, upper arm and sometimes underside, leg, tail, nape or back suffused with silvery gray or yellow; coronal crest variably developed. Formerly divided between *S. cristatus* and *S. phayrei*.

Dusky leaf monkey

Semnopithecus obscurus
Dusky or Spectacled leaf monkey.

Tripura (NE India), adjacent Bangladesh, N Shan and lowland SW and S Burma, Malay Peninsula and neighboring small islands (not Singapore). Forest, scrub plantations, gardens. HBL 16–27in; TL 22–34in; wt 9–24lb. Coat: dark gray to blackish-brown; nape paler, occasionally yellowish-white; back centerline usually paler and sometimes with orange sheen; elbow, legs and base of tail often paler than back, occasionally pale grayish-yellow; paws and brow black or blackish-brown; underside yellowish-brownish or blackish-gray, dark brown or occasionally pale orange; coronal crest usually present in NW; frontal whorl in Shan subspecies. Includes *S. phayrei* (part).

Capped leaf monkey

Semnopithecus pileatus
Capped or Bonneted leaf monkey.

Bangladesh and Assam E of Jamuna and Manas rivers, N and highland W Burma. Forest, swamp, bamboo. Overlaps with Dusky leaf monkey in Tripura and adjacent Bangladesh. HBL 19–30in; TL 32–43in; wt 20–31lb. Coat: upperparts gray, darkest on anterior of back and occasionally tinged with orange; paws and base of tail black or dark gray; paws sometimes partially orange-white; cheeks and underside gray, whitish to orange; crown hairs semi-erect and project over cheek hairs. Scrotum absent.

Golden leaf monkey [R]

Semnopithecus geei

Bhutan and Assam W of Manas river. Forest, plantations. HBL 19–28in; TL 28–37in; wt 21–26lb. Coat: orange-white; underside and sometimes cheeks, rear of back, orange; blackish hair tips on cap; faint gray tinge on forearm and shank, sometimes on rear of back and upperside of tail; crown hairs semi-erect and project over cheek hairs. Pubic skin pale. Scrotum absent.

Black colobus monkeys

Genus *Colobus*

Stomach 3-chambered; larynx large; sac below hyoid bone; facial skin black.

Satanic black colobus [V]

Colobus satanas

Macias Nguema, E and SW Cameroon, Rio Muni, NW Gabon and probably W Congo. Forest, meadows. HBL 23–28in; TL 24–38in; wt 13–24lb. Coat: entirely glossy black: crown hairs semi-erect and forward directed on brow.

White-epauletted black colobus

Colobus angolensis
White-epauletted or Angolan black or black-and-white colobus.

NE Angola, SW, C and NE Zaire, SW Uganda, W Rwanda, W Burundi, W and E Tanzania, coastal S Kenya, possibly Malawi, vagrant in NW Zambia. Forest; woodland maize cultivation. HBL 18–26in; TL 25–36in; wt 13–25lb. Coat: glossy black with white or whitish-gray cheeks, throat, long-haired shoulder epaulettes and tip or occasionally major part of tail;

CONTINUED ▶

narrow brow band sometimes, and chest region occasionally, white; brow fringe, frontal whorl or nape parting sometimes present.

Guinea forest black colobus

Colobus polykomos
Guinea forest or Regal or Ursine black colobus, or Western black-and-white colobus.

Guinea to W Nigeria, with hiatus at Dahomey Gap. Forest; scrub-woodland in Guinea savanna. HBL 22–27in; TL 28–39in; wt 13–26lb. Coat: glossy black; tail white; face border and throat sprinkled with white extending to long-haired shoulders, or wholly white which is absent or only sparsely sprinkled on shoulders; if absent is replaced by white outside to thigh; Point of nose reaches, or protrudes beyond, mouth.

Guereza

Colobus guereza
Guereza, or White-mantled or Magistrate black colobus or Eastern black-and-white colobus.

N Congo, E Gabon, Cameroon, E Nigeria, Central African Republic, NE Zaire, NW Rwanda, Uganda, S Sudan, Ethiopia, W Kenya and adjacent Tanzania. Forest, woodland, wooded grassland. HBL 18–27in; TL 20–35in; wt 12–32lb. Coat: glossy black; face and collosities surrounded by white; U-shaped white mantle of varying length on sides and rear of back; outside of thigh variably whitish; tail variably bushy and whitish or yellowish from tip towards base. Albinism common on Mt Kenya. Point of nose nearly touches mouth.

Red colobus monkeys

Genus *Procolobus*

Equatorial Africa. Limb proportions similar to *Pygathrix*; stomach 4-chambered; larynx small, no sac below hyoid. Sexual swelling in female and sometimes in immature male. Male skull usually with sagittal crest. Most, or all, placed in genus *Colobus* by some.

Guinea forest red colobus

Procolobus badius
Guinea forest red or Bay colobus.

Senegal, Gambia to SW Ghana. Forests, savanna woodland, savanna. HBL 18–25in; TL 20–29in; wt 12–22lb. Coat: crown, back, outside of upper arm and sometimes brow, outside of thigh and tail gray or blackish; pubic area white; rest of body whitish-orange to orange-red.

Cameroon red colobus [E]

Procolobus preussi
Cameroon or Preuss's red colobus.

W Cameroon. Lowland rain forest. HBL 22–25in; TL 29–30in; wt ? Coat: crown and back pale-stippled dark gray; cheeks, flanks, outside of limbs reddish-orange; tail blackish-red; underparts whitish-orange.

Pennant's red colobus

Procolobus pennantii

Macias Nguema, E Congo, W, N and E Zaire, SW Uganda, Rwanda, Burundi, Tanzania, Zanzibar. Forest. Exceptional polymorphism, especially in N Zaire, indicates more than one species may be involved. HBL 18–26in; TL 23–31in; wt 11–25lb. Coat: paws, crown, nape, anterior of back and tip of tail usually blackish-red or blackish-brown, sometimes paler or more orange; base of tail, rear of back blackish-brown to reddish-orange, occasionally black-flecked, and tail base occasionally orange-white below; flanks reddish-orange sometimes tinged with blackish, gray, brown or whitish or black-flecked, arm similar or whitish-black; leg reddish-orange, whitish-orange, gray or blackish-white, often tinged with brown; thigh brownish-black sometimes tinged with blackish, or yellowish; brow whitish, orange or reddish-orange to blackish-brown; cheeks and underside usually whitish or yellowish-white; cheeks and chest often tinged with orange. Whorl sometimes present, or behind ears. Subspecies include *P. p. kirkii*.

Tana River red colobus [E]

Procolobus rufomitratus

Lower Tana river, Kenya. Gallery forest. Dimensions not known. Coat: crown orange; back and tail dark gray; cheeks, limbs and paws pale gray; underparts yellowish-white. Whorl behind ears. Skull small.

Olive colobus [R]

Procolobus verus
Olive or Van Beneden's colobus.

Sierra Leone to SW Togo; C Nigeria, S of Benue River. Forest, abandoned cultivation. HBL 17–19in; TL 22–25in; wt 6–12lb. Coat: upperparts black-stippled grayish-orange, grayer towards limb extremities; pale gray below, grayish occasionally much reduced, so more orange above, and more whitish below. Short-haired longitudinal coronal crest flanked by whorl on either side.

► **Feeding on flowers** of white rhododendron, a Hanuman langur in the Himalayas.

Infanticide

Male takeovers in Hamuman langur troops

The Hanuman or Gray langur is the most terrestrial of all the colobines, and the most widespread primate other than man throughout the varied habitats of the Indian subcontinent and Sri Lanka. These elegant monkeys with steel-gray coats, black faces, hands and feet have long been considered sacred by the Hindu inhabitants of India, and Hanuman langurs may be found in close association with humans in villages and temples. The Hanuman langur is the largest of its genus of "leaf-eating monkey," measuring 24–30in (60–75cm) high when seated, with a tail up to 39in (100cm) long (the word langur derives from the Sanskrit for "having a long tail"). Fully adult males weigh around 40lb (18.4kg), females substantially less (24.5lb, (11.3kg), except in the Himalayan portion of the langurs' range where females weighing more than 35lb have been reported. Troops range in size from six to as many as 70 animals. The stable core of each troop is composed of female relatives who remain together in the same 90-acre or so (36-hectare) home range throughout their lives. Females as well as males play an important role in defending these feeding areas from exploitation by other troops. Use of the territory passes from mother to daughter.

Whereas females remain in the same location, young males leave the troop of their birth and join nomadic all-male bands containing from two to 60 or more males of various ages. These flexible assemblages travel over large areas, traversing the ranges of a number of troops, perpetually in search of breeding opportunities. Since most breeding troops contain only one fully adult male, competition between males for this position is fierce. Encroaching males are chased away from the troop. But occasionally one or more invaders will be successful at driving out the resident male and usurping his troop. In a number of cases, unweaned infants have been attacked, sometimes fatally, by incoming males. Some 32 takeovers by invading males have been reported, and at least half of these were accompanied by the disappearance of infants. Whether the occurrence or non-occurrence of assumed infanticide is due to genetic differences between males, local circumstances, or some interaction between the two remains unknown.

In areas like Mount Abu in Rajasthan, or Dharwar in Mysore state in south India, where food is plentiful, langurs live at densities up to 344 animals per square mile (133/sq km). Encounters between male bands and troops are frequent and takeovers common. Elsewhere, at sites such as Solu Khumbu and Melemchi high in the Himalayas, at the margin of this species' range, densities may be as low as 0.4 langur per square mile (1/sq km). Fighting among males does occur, but it is more subdued. Typically several adult males coexist in the same troop, and changes in male membership occur gradually over time. Only in the areas with frequent male takeovers (that is, where a takeover might be expected to occur on average once every two or three years) are male membership changes accompanied by assumed infanticide.

High takeover rates place males under considerable pressure to compress as much as possible of their harem's reproductive activity into their own brief tenure in the troop. Since mothers who lose their offspring become sexually receptive sooner than do mothers who rear their infants to weaning age, it has been suggested that the killing of infants by incoming males is an evolved reproductive strategy. While eliminating the offspring of his competitors, the usurping male enhances his own opportunities to breed. Infanticide in langurs has only been observed when males enter the troop from outside it. Similar patterns of male takeover accompanied by infanticide have recently been reported for the Sri Lankan Purple-faced leaf monkey and the Malaysian Silvered leaf monkey, as well as for more distant species of monkeys among the African cercopithecines, such as the Redtail

▲ **In the Himalayan foothills,** a Hanuman langur troop in Jammu and Kashmir state, northern India. Mixed troops have a stable core of related females and usually just one breeding male.

▼ **Border incident.** When two troops meet at the boundary between their ranges, both males and females join in defending their territory. Chases and hand-to-hand grappling look ferocious but animals are rarely injured in such encounters. It is after a male outsider has chased out a troop's breeding male that infants may be killed.

▲ **Before reaching maturity** males leave or are driven out of the troop and join other unattached males. These nomadic all-male bands monitor the strength of breeding troops, waiting for the opportunity for one of their number to usurp the resident adult male.

and Blue monkeys in the Kibale forest of Uganda, and in several species of howler monkeys in Central and South America. Because of the relationship between high population densities and the occurrence of infanticide, some have argued that infanticide may be a pathological behavior brought about through crowding. It is of special interest therefore that among howler monkeys and the cercopithecine monkeys in the Kibale forest, high population densities were not a factor.

The presence of infanticidal males creates peculiar challenges for females in the same population. However advantageous for the individual who increases his own reproductive success by eliminating an infant, infanticide is clearly disadvantageous from the point of view of the infant and its mother. Confronted with a population of males competing among themselves, often with adverse consequences for females, natural selection should have favored those females inclined and best able to protect their own interests. When an alien male approaches a troop, he is chased away by females as well as the resident male (or males). After a new male takes over, females may form temporary alliances to prevent him from killing infants. In the short term, females are often able to obstruct an infanticidal male, but the male has the option to try again and again, and often he eventually succeeds. (However, some males simply do not exhibit the infanticidal trait; others have been observed to make dozens of attacks on infants, yet never succeed.)

Given that the greater body size and strength of the male favor his eventual success, one of the most effective counter tactics may be a form of female deceit. If, as has been suggested, males are inhibited from attacking infants associated with former consorts, it may be because females already pregnant at the time of the takeover exhibit "pseudo-estrous" behavior (as they have been observed to do) and thus induce a male to tolerate her subsequent offspring by sexually soliciting a male at times when they could not ordinarily conceive. SB-H

APES

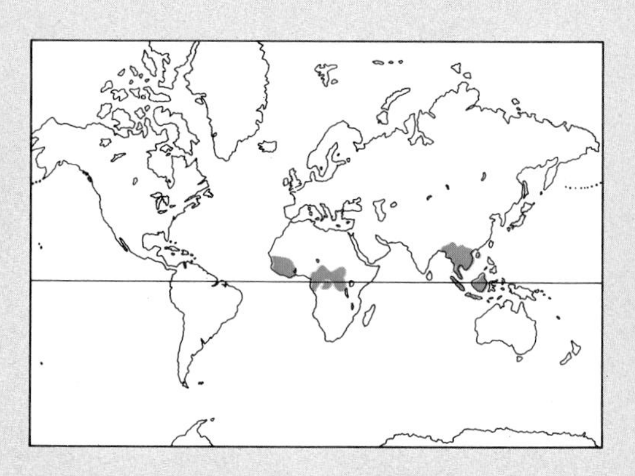

Families: Hylobatidae, Pongidae.
Thirteen species in 4 genera.
Distribution: E India to S China and south through SE Asia to Malay Peninsula, Sumatra, Java, Borneo, W and C Africa.

Habitat: chiefly rain forests; also deciduous woodland, savanna.

Gibbons or lesser apes (family Hylobatidae)
Nine species of the genus *Hylobates*, including **siamang** (*H. syndactylus*), **Hoolock gibbon** (*H. hoolock*), **Pileated gibbon** (*H. pileatus*), **Concolor gibbon** (*H. concolor*) and **Kloss gibbon** (*H. klossi*).

Great apes (family Pongidae)
Four species in 3 genera:
Common chimpanzee (*Pan troglodytes*) and **bonobo** or **Pygmy chimpanzee** (*P. paniscus*).
Orang-utan (*Pongo pygmaeus*).
Gorilla (*Gorilla gorilla*).

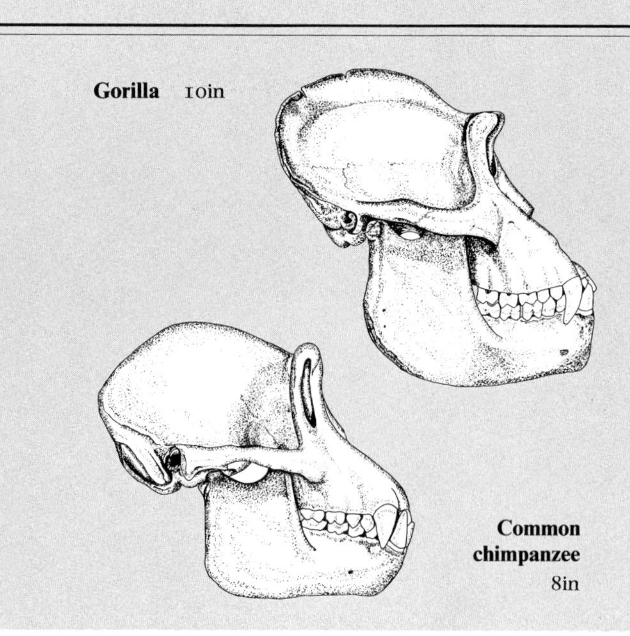

The apes are man's closest relatives, and with man comprise the primate superfamily Hominoidea. There is a fairly sharp distinction between the medium-sized lesser apes (gibbons and siamang, family Hylobatidae) and the considerably larger great apes (orang-utan, gorilla and chimpanzees, family Pongidae). The great apes and man are quite closely related and are the largest living primates. Fossil relatives of both the lesser apes and the great apes are known from the early Miocene, some 20 million years ago. Indeed the lesser apes may have become distinct as long ago as the early Oligocene (about 35 million years ago), but the fossil record is fragmentary.

The apes have no tail and their forelimbs, prominent in locomotion, are longer than the hindlimbs. The chest is barrel-shaped (rather than flattened from side to side as in monkeys) and the modified wrist structure permits greater mobility. Gibbons and siamang display the most spectacular pattern of movement, known as "true brachiation," in which the body is swung along beneath the branches with the arms taking hold alternately. The great apes, by contrast, are far less athletic in their movements through the trees. The orang-utan, the largest living arboreal mammal, moves slowly and deliberately, suspending the great weight of its body from all four limbs in a quadrumanous (four-handed) progression.

In contrast to their Asiatic relative, the great apes of Africa generally travel along the ground. Both chimpanzees and gorillas "knuckle-walk" with the knuckles of the hands providing the points of contact with the ground (orang-utans, particularly old heavy males, may descend to the ground and "fist-walk" on the outer margins of the hands—a more rudimentary pattern of forelimb support). Chimpanzees spend between a quarter and a third of their time in the trees, but gorillas only about 10 per cent.

The apes are all essentially vegetarian, though chimpanzees eat some animal food as well. Only the gorilla, the most terrestrial of the apes, is predominantly a leaf rather than a fruit-eater. In the remaining apes, particularly in gibbons and siamang, their "suspensory" movements while feeding are linked to adaptations for feeding in the terminal branches. Although animal food represents a small part of its diet, the chimpanzee's feeding activity has attracted much interest because of its possible relevance to the emergence of hunting behavior in early humans. At Gombe Stream in Tanzania, male chimps prey upon other primates (eg Red colobus and baboons) and other medium-sized mammals (eg Bush pig); the chimpanzees show a limited tendency to hunt cooperatively and they share food. Chimpanzees, mainly females, also feed on termites, using a "tool," a carefully selected twig (see p128).

Among the apes can be found almost every major pattern of social organization known among primates. The lesser apes are all monogamous and it appears that the mated pair remain together for life. The interval between births in gibbons and siamang is typically 2–3 years, sexual maturity is reached at about six years of age in both sexes, and sexually mature adults emigrate from the parents' home range. Consequently the maximum size of gibbon and siamang groups is about five individuals. Orang-utans feed in ones or twos, adult males typically avoid one another, and the only common social unit is the mother/offspring group, although adult

► **Face-flanges and throat sac** of the male orang-utan become swollen with fat in overweight captive individuals. These secondary sexual characteristics are acquired (like the silverback coloration of male gorillas) at 12–14 years, some 5 years after sexual maturity. By contrast the chimpanzees (males 15–20 percent heavier), siamang and gibbons show little or no size difference between the sexes and breed in monogamous pairs.

▼ **Size difference between sexes** is extreme in the orang and gorilla—mature males outweigh adult females almost twice over. In the tree-dwelling orang-utan BELOW this is all the more surprising since extreme sexual dimorphism is usually associated with a terrestrial life-style.

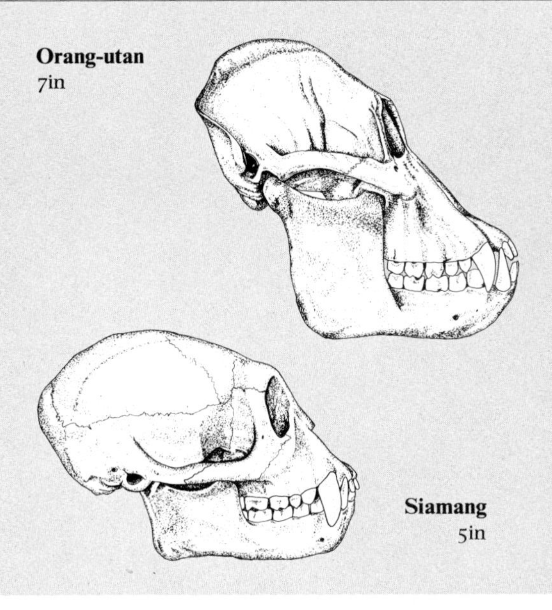

Skulls of Apes

Apes have relatively well-developed jaws, a flattened face with forward-pointing eyes and a globular brain-case, reflecting the fairly large brain in relation to body size. As in monkeys, each eye is enclosed in a complete bony socket formed by development of a plate of bone that separates the orbit from the jaw musculature behind it. The dental formula, as in Old World monkeys and man, is I2/2, C1/1, P2/2, M3/3 = 32, with spatulate (shovel-shaped) incisors and squared-off cheek teeth bearing relatively low cusps which reflect the predominance of plant food in the diet. The canine teeth are prominent and the rear edge of each upper canine hones against the front edge of the first lower ("sectorial") premolar. The lower jaw is fairly deep, and the two halves of the jaws are fused at the front as in monkeys and man. Similarly, in common with monkeys and man, the frontal bones of the skull are fused at the midline.

In association with their much smaller body size, the skulls of lesser apes (gibbons and siamang, family Hylobatidae) have relatively lightly built skulls, and there is relatively little difference between males and females (sexual dimorphism). In the much heavier great apes (orang-utan, chimpanzees, gorilla, family Pongidae), the skull is robustly built and sexual dimorphism in body size is reflected in the considerably heavier build of male skulls. Indeed, the powerfully developed jaw and neck muscles of male gorillas and orang-utans have required the development of substantial midline (sagittal) and neck (nuchal) crests to provide additional surfaces for the attachment of muscles. The orang's skull slopes back markedly more than skulls of other great apes.

females and/or immature orangs may form small temporary groups. Nevertheless, the orang-utan probably has some kind of social system based on overlapping ranges, perhaps broadly comparable to the "extended harems" of many nocturnal lemur and loris species (see pp24–41). The great apes of Africa have quite well-defined social groups. The gorilla is essentially harem-living, and groups average about 12 members consisting of a single mature silver-back male, a small number of younger ("black-back") males, several adult females, and immature animals. Lone silver-back males are also common. In chimpanzees there are at least two different levels of grouping. The fundamental social unit seems to be a community of 40–80 individuals including numerous adults of both sexes, but it is rare to find the whole community together in one place.

Clear-cut territorial behavior seems to be restricted to the lesser apes and the chimpanzees. Gibbons and siamangs are well known for their loud territorial calls (see p124). In chimpanzees territorial demarcation seems to be more discreet, though encounters between members of neighboring communities can lead to death.

The lesser apes, like most monogamous mammals, show very little difference in body size between sexes, though there are some differences in coat coloration between sexes. In chimpanzees there is relatively little size difference, but in both orang-utans and gorillas there is extreme sexual dimorphism in body size.

The apes, particularly the great apes, are very similar to human beings in their reproductive biology, and in all apes maternal care of infants is prolonged, ranging from some 18 months in gibbons to almost three years in great apes. RDM

GIBBONS

Family: Hylobatidae.
Nine species of the genus *Hylobates.*
Gibbons or Lesser apes.
Distribution: extreme E of India to far S of China, south through Bangladesh, Burma and Indochina to Malay Peninsula, Sumatra, W Java and Borneo.

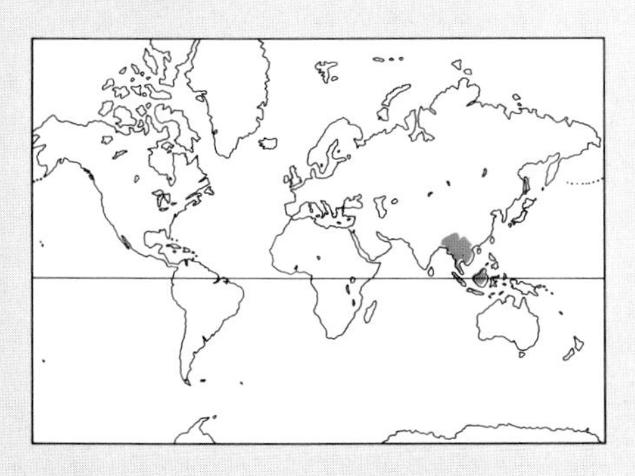

Habitat: evergreen rain forests (Southeast Asia) and semi-deciduous monsoon forests (mainland Asia).

Size: head-body length 18–26in (45–65cm) in most species, and weight 12.1–14.7lb (5.5–6.7kg). The siamang has a head-body length of 30–35in (75–90cm) and weighs some 23lb (10.5kg). Sexes similar in size.

Coat: color distinguishes species (and sometimes sexes and age group within a species), as do calls.

Gestation: 7–8 months.

Longevity: in wild, siamang to 25–30 years, Lar gibbon to 25 years or more.

► **Leaping to the next branch,** this Lar or White-handed gibbon shows the long arms, the thumb well distant from the grasping fingers, and broad feet with long opposable big toe of a true brachiator. Gibbons arm-swing (brachiate) beneath the branches, swinging the body along with alternate hand-holds.

▷ **"Suspensory" posture** of gibbons INSET is employed to reach ripe fruit in terminal branches as well as for locomotion.

The gibbons are distributed throughout the mainland and islands of Southeast Asia forming the Sunda Shelf and, being virtually confined to a life in the trees, depend on the evergreen tropical rain forests. It is the rapid clearance of these forests that places their future in such jeopardy. They have a spectacular arm-swinging form of locomotion (brachiation) and habitual erect posture, which are key adaptations for their unique suspensory behavior. They utter loud and complex calls of considerable purity in a stereotyped manner, which capture the spirit, both joyful and melancholic, of the jungles of the Far East. These beautiful calls, mainly given as duets, serve to develop and maintain pair bonds and to exclude neighboring groups from the territory of the monogamous family group. The gibbons' key attributes—monogamy, territoriality, a fruit-eating diet, "suspensory" behavior and elaborate songs—are a blend that is unique among primates.

Contrary to popular belief, the apes and man do not share an ancestor that habitually swung through the trees by its arms. While all apes have long arms, mobile shoulders and stand erect, only the gibbons have developed powerful propulsive abilities in their upper limbs.

While the great apes have developed sexual dimorphism in body size, adult male and female gibbons are more or less the same size. They are relatively small, slender and graceful apes with very long arms, longer legs than one would expect, and dense hair; they are more efficient at bipedal walking than the great apes, and do so on any firm support, such as branches too large to swing beneath, not just on the ground as is commonly supposed. Coat color and markings, especially on the face, clearly distinguish the species and, in some cases, age and sex. Some species have developed throat (laryngeal) sacs, which act as resonating chambers to enhance the carrying capacity of calls. These calls, especially those of the adult female, provide one of the easiest ways of identifying species.

In terms of diversity and abundance the gibbons are the most successful of the apes. From an adept climbing and fruit-eating ancestor, they have diversified throughout the forests of Southeast Asia over the last million or so years, maintaining the same body form and size (with the chief exception of the siamang) for hanging to feed from the terminal branches and for brachiating through the forest canopy. It was the frequent periods of isolation in different parts of the Sunda Shelf during the Pleistocene

► **Gibbon species are geographically separated,** except the siamang which overlaps both Lar and Agile gibbons. Within most species (not in the siamang, Kloss or Moloch gibbons) coat color varies according to sex and/or geographical population. (1) Siamang. (2) Concolor gibbon (**a**) black-cheeked and (**b**) white-cheeked phases. (3) Hoolock gibbon (4) Kloss gibbon. (5) Pileated gibbon. (6) Müller's gibbon. (7) Moloch or Silvery gibbon. (8) Agile gibbon (sexes similar in one population: (**a**), (**b**) forms in Malay Peninsula and southern Sumatra; and (**c**) southwest Borneo. (9) Lar gibbon (sexes similar in same population): Thailand, dark phase (**a**) and (**b**) light phase: (**c**) south of Malay Peninsula: (**d**) northern Sumatra.

Abbreviations: HBL = head-and-body length; TL = tail length; wt = weight.
E Endangered. V Vulnerable. I Threatened, but status indeterminate.

Siamang

Hylobates syndactylus

Malay Peninsula, Sumatra. HBL 29–35in; wt 23lb. Male, female and infants black; throat sac gray or pink. Calls: male screams; female—bark-series lasting about 18 seconds.

Concolor gibbon I

Hylobates concolor
Concolor, Crested or White-cheeked gibbon.

Laos, Vietnam, Hainan, S. China. HBL 18–25in (as all other gibbons); wt 12lb. Coat: male black with more or less whitish (or reddish) cheeks; female buff or golden sometimes with black patches; infant whitish. Calls: male grunts, squeals, whistles; female—rising notes and twitter, sequence of about 10 seconds.

Hoolock gibbon

Hylobates hoolock
Hoolock or White-browed gibbon.

Assam, Burma, Bangladesh. wt 12lb (female), 12lb (male). Coat: male black, female golden with darker cheeks, both with white eyebrows; infant whitish. Calls: male—di-phasic, accelerating, variable; female—similar to but lower than, and alternating with, male's.

Kloss gibbon V

Hylobates klossi
Kloss gibbon or beeloh (incorrectly Dwarf gibbon, Dwarf siamang).

Mentawai Islands, W Sumatra (Siberut, Sipora, N and S Pagai). wt 13lb. Coat: overall glossy black, in male, female and infant—the only gibbon so. Calls: male—quiver-hoot, moan; female—slow rise and fall, with intervening trill or "bubble" or not, sequence lasts 30–45 seconds.

Pileated gibbon E

Hylobates pileatus
Pileated or Capped gibbon.

SE Thailand, Kampuchea W of Mekong. wt ? Coat: male black with white hands, feet and head-ring; female silvery-gray with black chest, cheeks and cap; infant gray. Calls: male—abrupt notes, di-phasic with trill after female's; female—short, rising notes, rich bubble; 18 seconds.

Müller's gibbon

Hylobates muelleri
Müller's or Gray gibbon.

Borneo N of Kapuas, E of Barito rivers. wt ? Coat: mouse-gray to brown, cap and chest dark (more so in female), pale face-ring (often incomplete) in male. Calls: male—single hoots; female—as Pileated gibbon but notes shorter; sequence 10–15 seconds.

Moloch gibbon E

Hylobates moloch
Moloch or Silvery gibbon.

W Java. wt 13lb. Coat: silvery-gray in male and female, all ages; cap and chest darker. Calls: male—simple hoot; female—like Lar gibbon at first, ends with short bubble; 14 seconds.

Agile gibbon

Hylobates agilis

Malay Peninsula, Sumatra (most), SW Borneo. wt 13lb. Coat: variable (but same in both sexes in one population), light buff with gold, red or brown; or reds and browns; or brown or black; white eyebrows and cheeks in male, brows only in female. Calls: male—di-phasic hoots; female—shorter than Moloch gibbon, lighter-pitched, rising notes to stable climax, sequence lasts 15 seconds.

Lar gibbon

Hylobates lar
Lar or White-handed or Common gibbon.

Thailand, Malay Peninsula, N Sumatra. wt 12lb (female), 12lb (male). Coat: variable but same in both sexes in one population; Thailand: black or light buff; white face-ring, hands and feet. Malay Peninsula: dark brown to buff. Sumatra: brown to red, or buff. Calls: male—simple and quiver hoots; female—longer notes than Moloch gibbon, climax fluctuates, duration (Thailand) 18, (Malay Peninsula) 21, (Sumatra) 14–17 seconds.

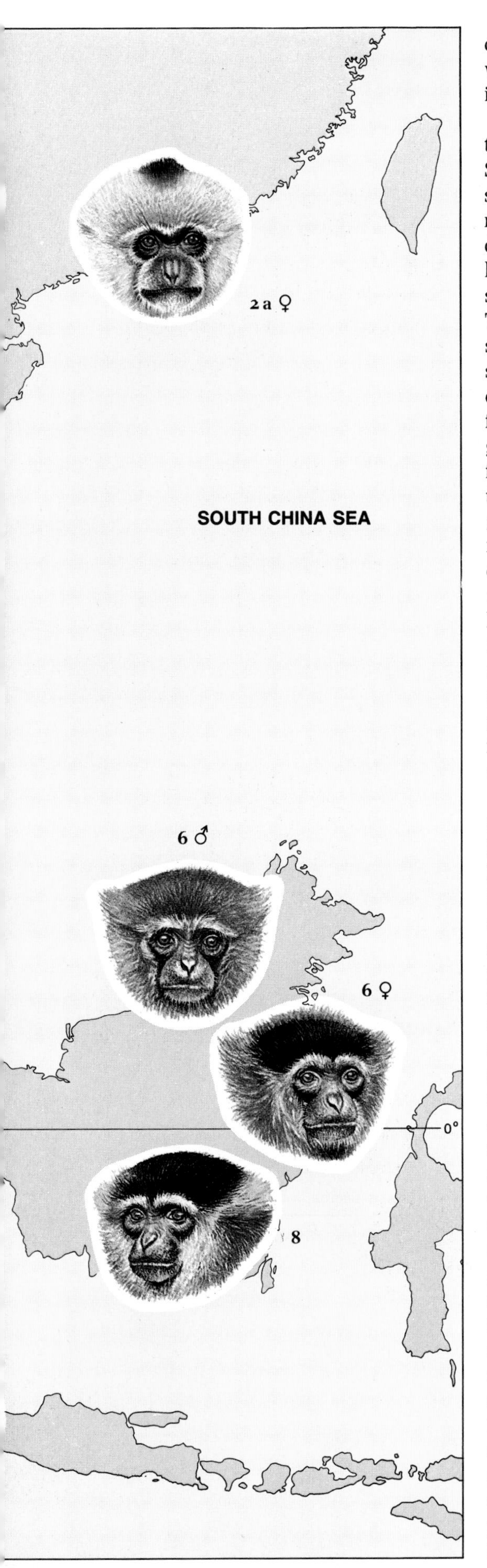

changes in sea level (7–2 million years ago) which stimulated differentiation of gibbons into present species.

It seems that about one million years ago the ancestral gibbon spread down into Southeast Asia to become isolated in the southwest, northeast and east (the Asian mainland would have been uninhabitable during the early glaciations). These three lineages respectively gave rise to the siamang, the Concolor gibbon, and the rest. The greatest changes occurred subsequently in the eastern group, which spread back towards the Asian mainland during the interglacial periods, giving rise first to the Hoolock gibbon (and the Kloss gibbon in the west), then to the Pileated, and finally, during and since the last glaciation to Agile and Lar, with Müller's and Moloch gibbons evolving on Borneo and Java, respectively. The range of gibbons has contracted southwards in historic time—1,000 years ago, according to Chinese literature, they extended north to the Yellow River. Curiously, gibbons are sexually dichromatic across the north of their range (males are mainly black, females buff or gray), black in the southwest, very variable in color in the center of their distribution, and tending to gray in the east.

The nine species are separated from each other by seas and rivers, except for the much larger siamang which is sympatric with the Lar gibbon in peninsular Malaysia and with the Agile gibbon in Sumatra. Although otherwise similar in size and shape, as a result of their common adaptation to a particular forest niche, they are readily identified by coat color and markings and by song structure and singing behavior. The siamang used to be placed in a separate genus, but the gibbons are best considered as monogeneric, with the siamang in one subgenus (*Symphalangus*) (50 chromosomes, diploid number) the Concolor gibbon of the northeast as a second (*Nomascus*), with 5–6 subspecies spread from north to south across the seas and rivers of Indochina (52 chromosomes), and the Lar gibbon group in a third (*Hylobates*) in the center and east (44 chromosomes). The concolor gibbon is as different from the Lar group as is the siamang. The Hoolock gibbon is the most distinctive of the third and largest subgenus, and it has now been found to have only 38 chromosomes; it is sexually dichromatic, like the Concolor and Pileated gibbons. The Kloss gibbon used to be called the Dwarf siamang because it is also completely black. The "gray" gibbons are the Moloch and Müller's gibbons. The most widely distributed and variable in color are the Agile gibbon, with at least two subspecies, and the Lar gibbon which can both be described as polychromatic, although the Lar gibbon in Thailand shows extreme dichromatism apparently not related to sex.

Gibbons generally show preferences for small scattered sources of pulpy fruit, which brings them into competition more with birds and squrrels than with other primates. Unlike the monkeys which feed in large groups and can more easily digest unripe fruit, gibbons eat mainly ripe fruit; they also eat significant quantities of young leaves and a small amount of invertebrates, an essential source of animal protein.

The structural complexity of the gibbons' habitat buffers the effects of any limited seasonality. Within as well as between plant species (climbers as well as free-standing trees) fruiting occurs at different times of year, ensuring year round availability of fruit. Since such plant species rely on animals for dispersal of seeds, this is an important example of co-evolution between plants and animals.

About 35 percent of the daily active period of 9–10 hours is spent feeding (and about 24 percent in travel). Feeding on fruit occupies about 65 percent of feeding time, on young leaves 30 percent, except for the siamang which eats 44 percent fruit and 45 percent leaves, and the Kloss gibbon (72 percent fruit, 25 percent animal matter and virtually no leaves). The larger the proportion of leaves in a species diet, the relatively larger are the cheek teeth and their shearing blades; the voluminous cecum and colon indicate an ability to cope with (and even ferment) the large leaf component of the diet in these simple-stomached animals. Fruit, even small ones, are picked by a precision grip of thumb against index finger, which permits unripe fruit to be allowed to ripen.

The adult pair of a gibbon family group usually produce a single offspring every 2–3 years, so that there are usually 2 immature animals in the group, but sometimes as many as 4. Thus, copulation is not seen very often; it is usually dorso-ventral with the female crouching on a branch and the male suspended behind, but occasionally the animals copulate facing each other. Gestation lasts 7–8 months and the infant is weaned early in its second year. The siamang is unusual in the high level of paternal care of the infant; the adult male takes over daily care of the infant at about one year of age, and it is from him that it gains independence of movement (by three years of age). Juveniles of either sex are

relatively little involved in group social interactions. By about six years the immature animal appears fully grown, and, as a subadult, tends to interact with siblings in a friendly manner, with the adult male in both friendly and aggressive ways, and avoids the adult female. Conflict with the adult male helps to ease the now socially mature animal out of the group by about eight years of age.

Subadult males often sing alone, apparently to attract a female, but they may also wander in search of one. Thus, either sons or daughters may end up near their parents, although sons perhaps more often. It is clear, however, that the first animal that comes along is not necessarily a suitable mate for life.

The siamang is unusual among gibbons in the high cohesion of the family group throughout daily activities—group members are 33ft (10m) apart on average, and rarely is one animal separated by more than 100ft (30m). In other gibbons the family feeds together only in the larger food sources; for the rest of the day they forage individually across a broad front of about 165ft (50m) coming together occasionally to rest and groom and, in some cases, to sleep at night.

Social interactions are infrequent; there are few visual or vocal signals, even in siamang, despite the "expressive" faces and complex vocal repertoire. Grooming is the most important social behavior, both between adults and subadults, and between adults and young; play, centered on the infant, is the next most common.

The most dramatic and energetically costly social behavior is singing, which mostly involves the adult pair. While it is most commonly explained as a means of communicating between family groups on matters of territorial advertisement and defense, there is increasing evidence that singing is crucial not only in forming a pair bond, but in maintaining and developing it. The elaborate duet which has evolved in most species is given for 15 minutes a day on average and from twice a day to once every five days, according to the species and to factors relating to fruiting, breeding and social change. The Kloss gibbon and possibly the Moloch gibbon do not have duets, but female Kloss gibbons have an astonishing "great call" (see p124) and male solos are given at or before dawn in Kloss, Lar, Agile, Pileated and perhaps Moloch gibbons.

This near-daily advertisement of the presence of a group and its determination to defend the area in which it resides is augmented by confrontations at the territorial boundary about once every five days, for an average 35 minutes. Altogether there are perhaps five levels of territorial defense—calls from the center, calls from the boundary, confrontation across the boundary, chases across the boundary by males, and, very rarely, physical contact between males.

The song bouts of most gibbon species conform to the same basic pattern: an introductory sequence while the male and female (and young) "warm up," followed by alternating sequences of organizing (behaviorally and vocally) between male and female, and of "great calls" by the female, usually with some vocal contribution from the male, at least at the end as a

▲ ◄ **In the Sumatran jungle** ABOVE, a siamang couple call in unison, the female (left) giving a series of barks while the male (center) screams. The song cycle lasts some 18 seconds while the subadult offspring (right) looks on. The throat-sac LEFT acts as a resonator to enhance the carrying quality of the call. Gibbon species can be distinguished by their songs, especially the great call of the female.

◄ **The Moloch or Silvery gibbon** TOP LEFT inhabits the western end of the island of Java. About twice as much time is spent feeding on ripe fruit, the staple diet, as on leaves. (The larger siamang eats more leaves than fruit.)

coda. Only in the Kloss gibbon are the songs of male and female completely separated into solos and these are discussed in the following pages. In Lar, Agile, Moloch, Müller's and Concolor gibbons, male and female contributions are integrated sequentially into the duet, whereas in Hoolock and Pileated gibbons and in the siamang, the male and female call together at the same time, even during the female's great call.

There are usually 4–8 gibbon family groups, each of 4 individuals, to each square mile (2.59sq km) of forest with a total body weight of 40–90lb per sq mi (45–100kg/sq km), but there may be less than one group or more than six. These groups travel about 1mi a day (siamang, Pileated and Müller's gibbons have mean annual day ranges of about 0.5mi) around a home range usually of 74–99 acres (30–40 hectares), but about 37 acres (15 hectares) for Lar in Thailand and for Moloch gibbons, and about 148 acres (60 hectares) for Lar gibbons in Malaysia sympatric with siamang. Most gibbon species defend about three-quarters of this home range (62 acres) as the group's territory. (About 90 percent of the home range is defended by Moloch and Müller's gibbons, and only about 60 percent by siamang and Kloss gibbons.) It is difficult to define territorial boundaries in siamang, however, since disputes are rare; it seems that they use their much louder calls to create a "buffer zone" between territories. Even though twice the size of other gibbons, siamang live in rather smaller home ranges, moving about less and eating more of more common foods such as leaves.

In 1975 there were estimated to be about 4 million gibbons. It was predicted that at prevailing rates of forest clearance (which still show little sign of abating) by 1990 these numbers would be reduced by 84 percent to about 600,000 with only the siamang, Müller's, Agile and Lar gibbons remaining in viable numbers. But thousands of gibbons (millions of animals in all) are still being displaced annually and die as a result.

The highest priority must be given to protecting adequate areas of suitable habitat for each species and subspecies. In the long term it would pay humans better, for economic reasons alone, to maintain rain forests rather than to clear them. As a further immediate step, displaced animals should be rounded up and either returned to unpopulated forests, if such schemes can be proved viable, or serve to establish breeding centers, preferably in the countries of origin, for research beneficial to the species and its conservation (eg nutrition and reproduction), for education and, if absolutely essential, for biomedical research of benefit to both humans and the captive maintenance of gibbons.

The gibbons hold a special place in the society of forest peoples, because of their resemblance to man (lack of tail, upright posture, intelligent expression). They tend not to be hunted by such people, but to be revered as a good spirit of the forest home. It is the more recent arrivals to these oriental countries who shoot anything that moves, but gibbons are very elusive to hunters. As forest dwellers they are neither pests nor effective carriers of disease. It is the human tendency to clear the jungles of the Far East which threatens the survival of every inhabitant.

DJC

Defense by Singing

Great calls and songs of the Kloss gibbon

Kloss gibbons are restricted to the Mentawai Islands west of Sumatra. Here the tropical rain forest contains several hundred potential food species, many of which flower and fruit in response to different cues. In such a complex environment it would be grossly inefficient for a monogamous group of primates such as these gibbons simply to wander the forest in search of food. The key to survival for a family group of Kloss gibbons is therefore to have an area of rain forest over which they have exclusive rights. Within its home range of 50–85 acres (20–35 hectares) the group can monitor the location and development of fruit, learn pathways between different points, and establish boundaries between its own and neighboring ranges.

A walk in the forest just before dawn gives a clue to how the Kloss gibbon defends its territorial rights. About two-thirds of the home range—the territory—receives particularly heavy use and all the trees used by the gibbons for sleeping at night lie within this area. Members of other groups will not be tolerated within the territory and the resident male will actively defend it if a

▲ **Launched into mid air** a female brings her great call to a climax, and serves notice on others to steer clear of her mate. She may also ABOVE run upright along branches tearing off leaves and be joined by other family group members shaking branches to enhance the display and giving calls of their own, like the clinging infant RIGHT.

second male comes too close. Such conflicts are rare, however, and this is due largely to the messages of passive defense that the adult males communicate when they sing. An adult male Kloss gibbon usually sings in the last hour before dawn and sometimes after dawn too. He will sing, on average, every other day and a song bout can last anything from 10 minutes up to two hours. It starts as simple whistling notes but, in longer bouts, is gradually elaborated into complex phrases containing about 12 notes and a trill. The male is relaxed when singing and often finds time to forage for insects or even eat fruits. His song's basic message is "I'm here," but, of course, it is hardly necessary to sing for two hours for his neighbors to understand just that. Comparative size is commonly used by animals to settle a territorial dispute instead of resorting to blows, but gibbon males in tropical rain forest could not be any larger because their feeding niche—that of exploiting the terminal branches—would then be unavailable. The very limited range of visibility also makes a show of size inappropriate. So, a male is able to "demonstrate" his confidence or willingness to defend his territory by the duration and complexity of his song. Unlike the simple "I'm here" message, this second message of a long, complex song need not be declared every day and makes it possible to minimize the time spent in conflicts with any particular neighbor.

All the trees he uses as platforms from which to sing are on or near the territory boundary and, since he uses a number of trees, his neighbors are able to get some impression of the area he is prepared to defend.

The song of the female Kloss gibbon has been described as "the finest music uttered by any land mammal" and comprises about twenty 30-second phrases consisting of long rising notes, a loud ringing trill and long falling notes. Her song, usually sung about two hours after dawn, is performed every third or fourth day, from large trees anywhere in the home range. The performance is one of the most dramatic events in the forest. She climbs to the upper boughs of a tall emergent tree to begin and at the climax of the "great call"—the trill—she launches herself into the air, swinging from bough to bough, tearing leaves off branches and frequently causing rotten branches to crash to the ground. During this climax the infant or juvenile will often cling to its mother's belly and attempt to sing as well, and the male and other juveniles race round shaking branches and generally making the performance as conspicuous as possible.

▲ **From a platform on the edge of his territory,** a male begins a bout of singing that may last up to two hours. He is answered TOP LEFT by his adult male neighbor on the adjoining mountain ridge. Meanwhile, supplementing a diet chiefly of fruits, the female searches for insects in foliage and the couple's offspring picks up ants on the back of its hand.

Unlike other gibbons, an adult pair of Kloss gibbons does not perform a vocal duet and the climax of the female's "great call" is essentially the only activity involving the whole group. While the female and her offspring interact in many ways, the male only rarely has contact with the other group members; he sleeps alone, and often travels around the home range some way behind or to the side of the female or offspring. The "great call" display is important therefore in maintaining the adult pair bond and the general cohesiveness of the group, but it is also a form of defense. Whereas the male's song is directed towards other males with designs on his home range, the female's song is directed towards other females and declares that any stray females who might have been attracted by the male's song should be aware that he already has a mate. Should a young female find an unattached, territory-holding male she will attempt to sing as soon as possible to stake her claim on him and his defended area.

Thus both sexes sing for defense—the male to defend his territory so that his mate and offspring can maintain themselves, and the female to defend her mate—the defender of the territory she occupies—from other females. AJW

CHIMPANZEES

The 2 species of the genus *Pan*.
Family: Pongidae.
Distribution: W and C Africa.

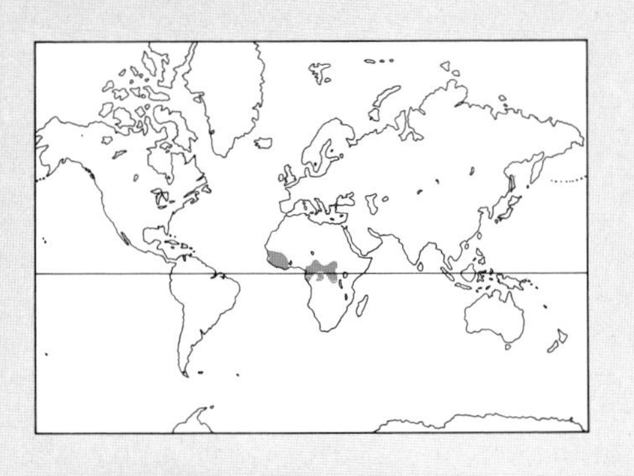

Common chimpanzee [V]

Pan troglodytes
Distribution: W and C Africa, north of River Zaïre, from Senegal to Tanzania.

Habitat: humid forest, deciduous woodland or mixed savanna; presence in open areas depends on access to evergreen, fruit-producing forest; sea level to 6,560ft (2,000m).

Size: head–body length 28–33in (70–85cm) (female), 30–36in (77–92cm) (male). No tail. Weights (poorly known in wild) in Tanzania 66lb (30kg) (female), 88lb (40kg) (male). Zoo weights up to 176lb (80kg) (female), 198lb (90kg) (male).

Coat: predominantly black, often gray on back after 20 years. Short white beard common in both sexes. Infants have white "tail-tuft" of hair, lost by early adulthood. Baldness frequent in adults, typically a triangle on forehead of males, more extensive in females. Skin of hands and feet black, face variable from pink to brown or black and normally darkening with age.

Gestation: 230–240 days.

Longevity: 40–45 years.

Pygmy chimpanzee [V]

Pan paniscus
Pygmy chimpanzee or bonobo
Distribution: C Africa, confined to Zaire between Rivers Zaïre and Kasai.

Habitat: humid forest only; below 4,925ft (1,500m).

Size: head-body length 28–30in (70–76cm) (female), 29–33in (73–83cm) (male). No tail. Weights (rarely measured in wild): 68lb (31kg) (female), 86lb (39kg) (male). Body lighter in build than Common chimpanzee, including narrower chest, longer limbs and smaller teeth.

Coat: as Common chimpanzee, but face wholly black, with hair on top of head projecting sideways. White "tail-tuft" commonly remains in adults.

Gestation: 230–240 days.

Longevity: unknown.

[V] Vulnerable.

Among the apes, it is the chimpanzees that can tell us most about the natural history of our common ancestors. In chimpanzee behavior we see many similarities to people—such as their tool-making, and their aggressive raiding parties of males—which show that several traits once thought to be uniquely human are not in fact so. Common chimpanzees live not only in humid closed-canopy forests but also in relatively dry areas, such as flat savanna where evergreen trees are confined to a few protected gullies. It was in open habitats such as these that our ancestors probably lived. Now, because of the impact of human activities on chimpanzee populations, the race is on to find out how chimpanzees live in their different habitats, before they or their habitats are destroyed.

Both species have stout bodies with backs sloping evenly down from shoulders to hips, a result of their relatively long arms (reaching just below the knee when standing erect). The top of the head is rounded or flattened (there is no sagittal crest), and the neck appears short. Their ears are large and projecting, while the nostrils are small and lie above jaws that project beyond the upper part of the face (prognathous muzzle). All their teeth are large compared to human teeth, but compared to gorillas chimpanzees have small molars, appropriate for their fruit diet. Bonobos, or Pygmy chimpanzees, have particularly small molars. Despite their name, however, their body size is not markedly different from Common chimpanzees.

Males are larger and stronger than females, and have bigger canine teeth which they use in severe fights with occasionally deadly results. Body proportions are otherwise similar, but both sexes have prominent genitals. Females in heat have prominent swellings of the pink perineal skin, lasting 2–3 weeks or more, every 4–6 weeks. Males have relatively enormous testes (4.2oz/120g).

Chimpanzees have similar sensory abilities to people; possibly they are better able to distinguish smells. Their large brains (18.3–24.4cu in/300–400cc) reflect a consistently high performance on all intelligence tests devised by humans, including the ability to learn and use words defined by hand signals in languages used by the deaf. In the wild, however, there is no evidence of linguistic abilities. Thirteen categories of chimpanzee calls have been recognized, from soft grunts given while feeding, to loud pant-hoots, consisting of shrieks and roars audible at least 0.6mi (1km) away.

Chimpanzees travel mostly on the ground, where they "knuckle-walk," like

gorillas. Like the other great apes, chimpanzees sleep in "nests," leafy beds normally made fresh each night. Adults sleep alone, infants with their mothers until the next sibling is born.

The Common chimpanzee, found north of the River Zaïre, has been known since the 17th century. There is much variation in body size and proportions, and even in coat and skin color. As many as 14 species were classified in the early years of this century, but only three subspecies are now recognized, in western, central and eastern populations. However, no distinctive traits have been established for the three subspecies, whose validity remains uncertain. Bonobos were first described as a separate species from museum collections in 1929. Their restriction to closed-canopy forests south of the River Zaïre has protected them from most collectors, and little is known in detail of their distribution or variation.

In all habitats chimpanzee diets are composed mainly of ripe fruits, which they eat for at least four hours a day. One to two hours daily are spent eating young leaves, particularly in the late afternoon. In long dry seasons tree seeds partly replace fruit, and flowers, soft pith, galls, resin and bark are also taken. Chimpanzees eat from as many as 20 plant species a day and 300 species in a year. Whenever possible they eat large "meals" from single food sources, which allows them to rest for an hour or two before walking on to the next fruit tree. They do not store food, and almost all food is eaten where it is found.

Animal prey make up as much as 5 percent of the diet by feeding time. Social

▲ **Adult chimp's black face** and ears are usually pink in younger animals.

◄ **Riding on the mother's back** begins at about six months of age and continues for several years. Earlier the single offspring clings to its mother's underside from within a few days of birth.

► **A pygmy chimpanzee or bonobo** in captivity. Despite its name this species is no smaller than its relatives on the northern banks of the River Zaïre.

Tool Use

Chimpanzees use tools to solve a greater range of problems, both in the wild and captivity, than any animal apart from humans. Two kinds of food are commonly obtained with tools, though chimpanzee populations vary in their use of them.

Most social insects have potent defenses which are overcome by the use of sticks or soft stems. For instance, chimpanzees prepare smooth, strong wands 24–28in (60–70cm) long for feeding on Driver ants: they lower the wand onto an open nest, wait for ants to crawl up it, then sweep them off and into their mouths before the ants have time to bite. They strip grass stems to make them supple, poking them into holes on termite mounds: soldiers bite the stem, and cling on long enough for the chimpanzee to extract and eat them. Sticks are also used to enlarge holes, so that honey or tree-dwelling ants can be reached.

A second food type eaten with tools is fruits with shells too hard to bite open. Sticks or rocks weighing up to 3.3lb (1.5kg) are used to smash these fruits, sometimes against a platform stone. Platform stones have been found with a worn, rounded depression, suggesting they have been used for centuries.

Tools are not used only when feeding. Adult males elaborate their charging displays by hurling sticks, branches or rocks of 8.8lb (4kg) or more: in a long display as many as 100 rocks are thrown, and other individuals have to watch out to avoid being hit. Missiles have once been seen in the context of hunting: a male hit an adult pig from 16ft (5m), startling it so that it ran off and allowed the chimpanzee to seize its young.

Fly-whisks, sponges (of chewed bark or leaves), leaf-rags and other tools are used irregularly in different areas. Why do chimpanzees use, and even make, so many types? Their upright sitting position, opposable thumb and precision grip are only part of the answer: other primates have them too. Nor are they unique in having large brains: gorillas and orang-utans are large-brained, but rarely use tools. Chimpanzees are inventive problem-solvers, and their habits may simply offer more tool-using opportunities than those of other great apes.

insects provide the largest amounts and are collected either by hand (eg aggregating caterpillars) or with tools (see LEFT). Females spend twice as much time as males eating insects. Birds are caught only occasionally, but mammals are regular prey in some areas and are known to be eaten particularly by males, wherever there have been long-term studies of chimpanzees. Monkeys, pigs and antelope are the principal prey, especially young animals. Hunting occurs irregularly, typically only when prey is surprised in appropriate circumstances. Monkeys, for example, are ignored in continuous canopy forest but are chased if encountered in broken canopy, with few escape routes.

Feeding is essentially an individual activity, and during periods of food scarcity most chimpanzees travel alone or in small family groups (mother with one or two offspring). Larger parties, often formed when individuals meet at food trees, give no advantages in obtaining food and can sometimes lead to competition for feeding sites. Even when hunting mammalian prey, chimpanzees show little cooperation; more than one chimpanzee may take part in the chase, but once a kill is made there is intense competition for the carcass. Both at kills and at trees with a few large fruit individuals with food in their hands are surrounded by others trying to take morsels. This leads to "food-sharing," where scrounging is tolerated apparently because it gives the possessor some peace.

Females raised in captivity begin mating at 8–9 years and give birth for the first time at 10–11 years. Wild females mature 3–4 years later. There is no breeding season. Common chimpanzee females are not receptive for 3–4 years after giving birth, then resume sexual activity for 1–6 months until conception. Female bonobos continue sexual activity during much of pregnancy and lactation. A single young is born after a gestation of 8–9½ months: twins are rare.

The newborn chimpanzee is helpless, with only a weak grasping reflex and needing support from the mother's hand during travel. Within a few days it clings to the mother's underside without assistance and begins riding on her back at 5–7 months. By 4 years the infant travels mostly by walking, but stays with its mother until at least 5–7 years old. Weaning occurs well before this, starting in the third year. Mothers groom and play with their offspring, and allow younger juveniles to take from them foods which only adults can get easily.

Males show precocious sexual behavior,

▲ **Two young chimps at play** while their mothers sit nearby. Mother and young continue to travel together for years after weaning and the bonds last into adulthood, reached at 13–15 years of age.

▼ **Rehabilitation of chimps** displaced by logging in the Gambia. A slow rate of reproduction makes chimpanzee populations especially vulnerable to destruction of their habitat. Rehabilitation centers reintroduce captured chimps to the wild.

including full intromission of adult females by the age of 2. Complete courtship patterns, however, develop slowly, from 3–4 years onward. They are normally directed towards females in heat (estrus), and consist of nonvocal attention-getting displays by the male sitting with penis erect: displays include hair erection, branch-shaking and leaf-stripping, some of this behavior varying between populations. Females respond by approaching and presenting for mating. Juveniles of both sexes commonly follow the female, and approach and touch the male after mating begins. These interfering juveniles make submissive gestures to the mating male, who responds with aggression only to older juvenile males.

Females mate only when in heat (when their sexual skin is swollen) and for the first week or more female Common chimpanzees are promiscuous and mate on average six times a day. Towards the last week of estrus, near the time of her ovulation, high-ranking males compete for mating rights, by threatening or attacking subordinates who approach the female. Alternatively, an exclusive "consortship" is formed, a female and male eluding other members of the community for days or weeks. Though 75 percent of matings occur in the promiscuous phase, pregnancies are most likely to result from consortships. The sexual behavior of bonobos is less well known. They are often promiscuous and, unlike Common chimpanzees, sometimes mate from the front.

All chimpanzee individuals are members of a community 15–120 strong, and travel sometimes with other community members. Neighboring communities have partially overlapping ranges, within which they show either resident or migratory traveling patterns. Resident communities occupy ranges of 4–20sq mi (10–50sq km), and have population densities of 0.4–2/sq mi (1–5/sq km). Within these community ranges individuals have their own dispersed "core" areas and spend 80 percent of their time within 0.8–1.5sq mi (2–4sq km). Common chimpanzee mothers often travel alone except for their offspring, covering 1.2–1.9mi (2–3km) per day. Males are more gregarious, traveling further than mothers and being attracted both to other males and to females in heat.

Party sizes average 3–6 for Common chimpanzees and 6–15 for bonobos. Parties of bonobos tend to have equal sex ratios and persist for longer than those of Common chimpanzees.

Less is known about migratory communities, which occur in open habitats in the extreme west and east of the chimpanzee's range, where population density falls to 0.02–0.4/sq mi (0.5–1.00/sq km). Communities migrate within large ranges of 77–154sq mi (200–400sq km) or more, settling for a few weeks in areas where food is temporarily abundant. When they settle they appear to behave like a resident community, with females more dispersed and more often alone than males.

Membership of communities is determined by sex. Males seldom or never leave the community where they are born, whereas most females migrate to a new community during an adolescent estrous period. For the young female a successful transfer occurs when she establishes her own core area within the range of another community where she can feed without aggression from other females. She travels in the company of males for much of the time in her first few months in a new community, and thereby obtains protection from hostile females. Females typically make a successful transfer only once, but repeated movements are known.

Within communities relationships between established females are poorly understood, but known to range from aggressive to friendly. Female bonobos associate more than Common chimpanzees, and rub their genital areas together in contexts of social excitement. Male relationships are more overt; tension is routinely expressed in dominance interactions when parties meet (see RIGHT), and males also spend much time grooming each other. Males tend to associate with their maternal brothers, but close associations between other male pairs are also common.

There are an estimated 50,000–200,000 chimpanzees and bonobos in 15 countries. In a further nine countries they were known in historical times but are now probably extinct. The main threat comes from habitat destruction, particularly commercial logging, eg in Ivory Coast and central Zaire. Commercial exploitation for overseas trade and hunting for bushmeat have severely reduced populations in some areas, eg Sierra Leone and eastern Zaire, respectively. About 10 national parks contain chimpanzees but most are small. Given their low reproductive rate, chimpanzees are highly vulnerable to loss of habitat or populations. Common chimpanzees are abundant in captivity but bonobos have no viable breeding populations outside the wild.

In several areas chimpanzees are protected by local custom (eg parts of Guinea, Zaire, Tanzania) and they visit fields or even markets. RWW

Gang Attacks

Cooperative fighting in chimpanzee society

Male chimpanzees are commonly aggressive to each other, in common with the males of many other species. But they are unusual because for much of the time adult males not only tolerate male company but seek out, groom and follow each other. This mixture of hostility and relaxed association occurs because familiar males form alliances and support each other in conflicts. Since the composition of male parties shifts unpredictably at any time, the presence of particular allies is never certain. Social relationships are therefore complex.

Males form a loose dominance hierarchy. Subordinates greet dominants by bobbing or crouching in front of them, giving "pant-grunt" vocalizations. Dominant males commonly approach others with a charging display, running with hair erect, perhaps dragging a branch. Reunions between familiar males are the principal context of these displays, which test whether any change has occurred in the dominant's confidence or the subordinate's willingness to challenge. Occasionally, of course, a reversal occurs, normally after several weeks of tension. Chases, physical attacks (hitting, rolling on the ground, stamping) and screams are frequent.

Displays are not restricted to such contexts of social uncertainty. They can also be prompted by heavy rain or strong winds, or by coming to a stream or waterfall. These displays are given both by lone males and by parties. The male begins by rocking gently where he sits, perhaps rhythmically swaying a branch. Movements become larger and more exaggerated until he swaggers or charges, on the ground or through trees and vines, sometimes at a slow tempo for up to 15 minutes. Wind, rain and rushing water all interrupt the quiet of the normal environment, but why chimpanzees respond with displays is unknown.

The importance of other displays is clearer. Given primarily by males, they lead to reduced tension because one male has acknowledged the other's competitive superiority. After reunions a typical sequence is for males to groom each other for several minutes before traveling together. If subsequently any competition is initiated, such as for mating rights, the subordinate abandons his challenge quickly.

But why do males travel together? Though they sometimes follow the same receptive female, or exploit the same productive food trees, they may also travel as an all-male party when food is abundant throughout the community's range. At such times the air is often full of long-distance calls as individuals gauge where others are and join to form large parties, 8–40 strong. This is only possible when food is abundant but provided that it is, chimpanzees (especially males) tend to congregate.

Male congregation is particularly striking on the borders of community ranges, and functions in both defense and attack. When parties from different communities meet by chance they give calls which reveal their relative size: smaller parties retreat faster. Provided there are several males they are not attacked. However, such interactions can lead to the loss of preferred feeding areas, and seasonal changes in food

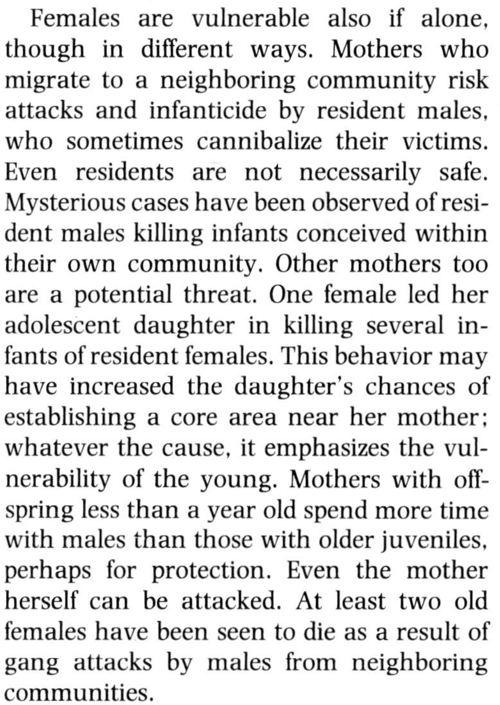

▲ **Aggression and coalitions** between male chimps. (1) Display: a male chimp swaggers bipedally, begins a charge beating its chest, then stops to stamp on and slap the ground, picks up and charges with a stick. (2) A subordinate male bobs and "pant-grunts" in front of another's dominant status, while in the background two other males engage in the mutual grooming which may follow such an encounter. (3) A "border patrol" turns on a lone male, then chases him and finally beats and bites him in a joint effort.

distribution mean that small communities may be regularly jostled by their neighbors. Even more important, parties meet not only by chance but also by design. Common chimpanzees form "border patrols," male parties which visit the boundary of their range and sit for up to two hours, listening and looking toward neighboring areas. If nothing is heard or seen, the patrol may return or advance. If a large party is detected calls are exchanged. And if a lone male is seen he may be stalked, chased, and attacked in a cooperative effort: some males hold him down while others hit or bite him. The injuries inflicted can lead to death within a few days. These are rare interactions, of course, but more than 10 deaths from such raids have been seen or suspected in relationships involving five different communities. If lone males are so vulnerable to attack by a larger party it is not surprising that males seek each other's company, and that dominant males tolerate the presence of subordinates.

Females are vulnerable also if alone, though in different ways. Mothers who migrate to a neighboring community risk attacks and infanticide by resident males, who sometimes cannibalize their victims. Even residents are not necessarily safe. Mysterious cases have been observed of resident males killing infants conceived within their own community. Other mothers too are a potential threat. One female led her adolescent daughter in killing several infants of resident females. This behavior may have increased the daughter's chances of establishing a core area near her mother; whatever the cause, it emphasizes the vulnerability of the young. Mothers with offspring less than a year old spend more time with males than those with older juveniles, perhaps for protection. Even the mother herself can be attacked. At least two old females have been seen to die as a result of gang attacks by males from neighboring communities.

The brutality and effort put into gang attacks implies they have important rewards. In one case a community with a few females lost all its males in a four-year period. Individuals from neighboring communities occupied the almost empty range, and in subsequent years both flourished. Whether competition for feeding space or mating rights is more important remains to be discovered.

ORANG-UTAN

Pongo pygmaeus E
Sole member of genus.
Family: Pongidae.
Distribution: N Sumatra and most of lowland Borneo.

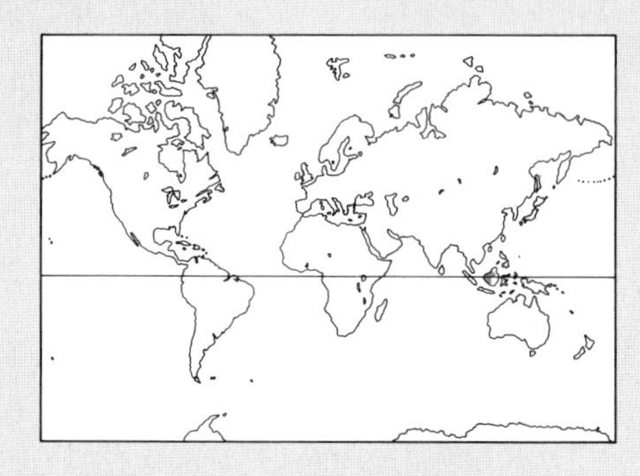

Habitat: lowland and hilly tropical rain forest, including dipterocarp and peat-swamp forest.

Size: head-body length (male) 38in (97cm); female 31in (78cm); height (male) 54in (137cm); female 45in (115cm); weight (male) 130–200lb (60–90kg); female 88–110lb (40–50kg).

Coat: sparse, long, coarse red hair, ranging from bright orange in young animals to maroon or dark chocolate in some adults. Face bare and black, but pinkish on muzzle and around eyes of young animals.

Gestation: 260–270 days.

Longevity: up to about 35 years (to 50 in captivity).

Subspecies: 2. *Pongo p. pygmaeus* from Borneo; *P. p. abelii* from Sumatra, thinner, longer faced, longer haired and paler colored than *P. p. pygmaeus*.

E Endangered.

► **Mother and young** ABOVE, one of perhaps three or four the female orang raises on her own during 20 years of active reproductive life in the wild.

▷ **The only truly arboreal ape,** the orang-utan swings on its long arms, grasping branches by its hook-like hands and feet in search of its chief food, fruit. The male's impressive size, cheek flaps, throat pouch and long hair enhance his display in confrontations with males that attempt to encroach too far in his home range.

The shy orang-utan or "Man of the Woods" is Asia's mysterious great red ape. It lives in the remote steamy jungles of Borneo and Sumatra and for a long time was known more from fabulous native stories and for his reputation as an abductor of pretty girls than from documented scientific accounts. In recent years, however, several extensive studies of this fascinating animal have been made and although some mysteries still remain the orang-utan is now one of the better known primates.

The orang-utan is a large, red, long-haired ape of very striking appearance. It is active in the daytime and spends most of its time in the trees. Adult males are about twice the size of females and have cheek flanges of fibrous tissue that enlarge their face and very long hair which enhances their aggressive displays. Orang-utans have very long arms for arboreal locomotion and hook-shaped hands and feet. They use their heavy weight to swing trees back and forth until they can reach across gaps between one tree and the next. Most orang-utans only occasionally descend from the trees to travel on the forest floor, though large males do so more. On the ground they use their strong arms to bend and break branches to obtain food and to make sleeping nests for the night. Their teeth and jaws are relatively massive for tearing open and grinding coarse vegetation, spiny fruit shells, hard nuts and tree bark. Orang-utans have a large throat pouch, most fully developed in adult males, which is inflated during calling and adds resonance to vocalizations, particularly the territorial "long call" of the adult male (see p134). The orang-utan has a large brain and is as highly intelligent as the other great apes.

Orang-utans are clearly descended from one of the Miocene *Sivapithecus* fossil apes but their exact ancestry is not known. Pleistocene orang-utans of giant size are known from China and subfossil orang-utans about 30 percent larger than present size are known from caves in Sumatra and Borneo. During the Pleistocene, a small form occurred on Java but is now extinct. It would seem that ancestral orang-utans were more terrestrial than those of today but the degree of anatomical and functional adaptation of orang-utans to tropical rain forest suggests a very ancient co-evolution. Biochemical relationships indicate that the orang-utan is a rather more distant relative of man than are the African apes, the gorilla and chimpanzee.

The orang-utan has a huge capacity for food and will sometimes spend a whole day sitting in a single fruit tree, gorging. About 60 percent of all food eaten is fruit, including such well-known tropical species as durians, rambutans, jackfruits, lychees, mangosteens, mangoes and figs. The remainder of the diet is mostly young leaves and shoots, but they regularly eat insects, mineral-rich soil, tree bark and woody lianas and occasionally eggs and small vertebrates, raiding nests of birds and squirrels. Water is drunk from tree holes; the ape dips in a hand and sucks the water-drops falling from its hairy wrist.

Orang-utans range slowly but widely, and usually singly, in search of fruit and they have an uncanny ability to locate it. In the efficiency of their travel routes they show great knowledge of the forest, its seasons, and the relative positions of individual trees and they can deduce where food is from observing the movements of other animals, especially hornbills and pigeons, which share items of diet. They find travel hard work and only travel a few hundred yards a day through the trees.

Orang-utans are long-lived, slow-breeding animals. Their breeding strategy is based on producing a few high-quality, well-cared-for young rather than mass production with high mortality. Females become sexually mature at about 10 years and consort for several days at a time with adult males during periods of sexual receptivity over several months until they become pregnant. They then live alone to bear and rear their infants, which are not weaned for three years. Infants ride on the mother's body and sleep in her nests until she has another infant. The interval between births can be as little as three years, but in fact in most wild populations adult females average only one infant every six years and remain fertile until about 30 years of age. The adult male's reproductive strategy consists of developing a range which takes in as many sexually responsive females as possible. He consorts with these females when they are receptive and until they are pregnant and is aggressive towards other adult males encroaching on his range. Once the females within his range are pregnant or caring for infants, they are sexually uninteresting for several years, so high-dominance males then sometimes move to another area of activity to include more females.

Orang-utans are rather solitary animals. Apart from occasional sexual consortships, adult animals travel and forage independently, each animal occupying an individual though non-exclusive home range of several square miles. Infants remain with their mothers. Juveniles (3–7 years old) become increasingly independent, sometimes traveling alone, and by adolescence (7–10 years old) they have usually completely left their mothers. Juveniles and adolescents are the most social, as in other apes, and sometimes come together to play for a few hours or even travel around in pairs or tag onto family units. When several adult orang-utans meet, as when attracted to the same major food source such as a fruiting fig tree, they show almost no social interaction and depart separately when they have eaten their fill. It is nevertheless clear that despite this apparent lack of interest in each other wild orang-utans do recognize individually all the animals whose ranges they regularly overlap, and they have a good knowledge of the location of other individuals nearby in the forest. Subadult males (10–15 years of age) are mainly solitary, do not call and sometimes secretly consort with females.

Adult males in particular are very conscious of one another's movements and give loud "long call" vocalizations to advertize their whereabouts. Encounters between males are avoided but when they occasionally do meet they indulge in violently aggressive displays when they stare at each other, inflate their pouches or charge about, shaking and breaking branches and sometimes calling. Usually one male backs down and flees along the ground but occasionally fights occur and antagonists grab and bite their rivals. Most adult males carry the scars, cut facial flanges or stiff broken fingers, from past battles. Orang-utan populations which are becoming condensed and overcrowded, or otherwise socially disturbed, as a result of habitat loss to logging operations, exhibit increased levels of aggression between males, reduced level of consortship and reduced reproductive rate. Such changes may be adaptive in enabling populations to restabilize with minimal conflict and minimal risk to infants.

There is also evidence that adult males continue their aggressive territorial behavior even after they have ceased to be sexually active after about 30 years of age. This suggests that longevity has been selected to enable males to defend space until their eldest offspring are old enough to take over the same space. More long-term field data are needed to test these theories.

Man has been a serious competitor of the orang-utan since his ancestors moved into the Asian rain forest over 9 million years ago. Orang-utans prefer just those fruits that man also enjoys eating, and man has consistently destroyed the apes' forests and even hunted orang-utans for food.

The Dayak and Punan people of Borneo show orang-utans great respect. Some feel spiritually related to orang-utans, others think of orang-utans as descended from a disgraced man who fled into the forest. Many native stories concern sexual relationships between man and ape and woman and ape. Indeed captive orang-utans are extremely precocious sexually, and pet orang-utans were sometimes used for ribald games at the end of a good longhouse party.

Orang-utans make sensitive, gentle pets

The Long Call of the Male

From time to time the peace of the jungle morning is broken by a curious noise: a loud crash as a weak branch is broken off and hurled to the ground, followed by a series of loud roars which rise to a crescendo of bellowing then die back to repetitive bubbling groans. The whole sequence may last for a minute or two; this is the male orang-utan's long call.

According to native legend it is the male orang-utan expressing his anguish for the loss of his human bride when she escaped from the treetop nest that was her prison. Scientists still argue as to the call's exact function. Is it territorial to drive away other males, a primarily sexual display to attract receptive females, or a social signal to inform the whole community of the whereabouts of the patriarch? Probably the call serves all these functions. Observations on calling males show there is an increased tendency to call in bad weather, when other males are calling, when the caller meets other males or when he is close to sexually receptive females.

The response of orang-utans on hearing the call varies. Most give no visible reaction. Intruding or subadult males tend to move away quietly whereas other dominant males call in reply and advance to challenge the caller. Some females with young hide in the top of a tree if they hear a calling male near by; other females are attracted to the caller and consort with him.

It seems clear that the long call does have a spacing function among adult males at least. Large males keep their distance and calls seem a likely way for them to monitor one another's movements. Long calls probably attract receptive females and may also act as a coordinating signal for the whole orang-utan population in the area. Coordinated seasonal movements of whole communities over several miles have been noted and could be guided by the males' calls.

◀ **Young orangs differ from adults** in the upright hairs on the crown and pink coloration on the muzzle and round the eyes. In adults the darker coat is flat on top of the head and the facial skin is all black.

▼ **Young Bornean orang-utan** (the individual opposite is from Sumatra). A quadrupedal climber, the orang progresses slowly through the trees, using its body weight to bend trunks or branches in the desired direction.

and were very popular in that role before it was made illegal in Malaysia and Indonesia to own or sell one. Captive males sometimes became rather bad-tempered and some apparently docile animals have turned viciously on their keepers, biting off fingers etc.

Captive orang-utans score very highly in comparative intelligence experiments, which is somewhat surprising in view of their relatively simple life-style and social relationships. The orang-utan's high intelligence is probably related to its extraordinary ability to find fruit in the tropical rain forest, where fruit is usually scarce and on isolated trees. A good memory for time and place and an ability to make deductions are essential in predicting the whereabouts of food.

Since they are closely related to man, orang-utans are of interest as carriers of several human diseases (such as malaria, other blood fevers and viral infections) and parasites but contact in the wild is so rare that this is not considered a serious health problem.

Great concern has been shown for the plight of the orang-utan. This spectacular ape has already vanished from several of its former haunts and its home, the tropical rain forest, is disappearing at a frightening rate due to logging for timber and land clearance for agriculture.

At one time the species was seriously threatened by the trade in orang-utan babies for zoos and as pets. Mothers were shot to capture their young and many young died during capture and transport. The elimination of this dreadful trade and improvements in the protected status of the animal have greatly relieved the situation.

Research has shown that orang-utans are not as rare as was formerly thought—density varies between 0.4 and 2 animals per square mile (1–5/sq km) depending on habitat quality—but the future of the rain forest remains as uncertain as ever. Eventually, more rational utilization and management of the rain forests may come about, but in the short term the only way to save the orang-utan is to protect as much of its habitat as possible within the boundaries of nature reserves and national parks. Fortunately, conservationists in Indonesia and Malaysia have been very successful in establishing such reserves, and major populations are now protected in the Gunung Leuser National Park in Sumatra, Tandjung Puting and Kutai National Parks and Gunung Palung and Bukit Raja reserves in Kalimatan (Indonesian Borneo), Lanjak Entimau reserve in Sarawak and the proposed Danum Valley reserve in Sabah.

Several rehabilitation stations have been established in Malaysia and Indonesia to train confiscated young pet orang-utans to return to the wild. Some success has been achieved, particularly in drawing local attention to the plight of this superb ape. It has, however, been generally agreed that returning human-oriented animals, capable of carrying human diseases, into healthy wild populations is not a useful exercise. Instead rehabilitated animals should be released in areas where wild orang-utans no longer occur, in order to establish new populations. JMacK

GORILLA

Gorilla gorilla [V]
Sole member of genus.
Family: Pongidae.
Distribution: C Africa.

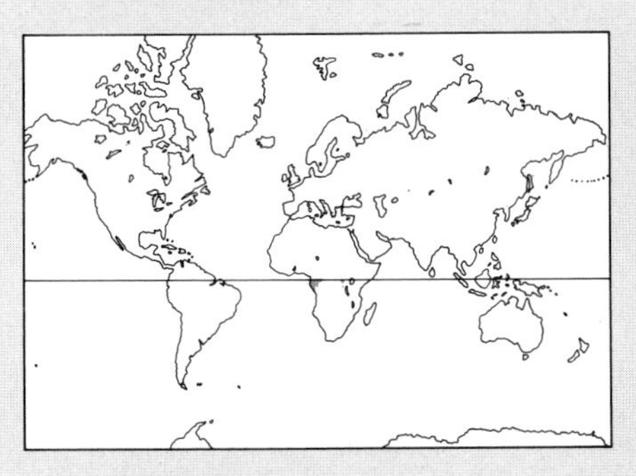

Habitat: tropical secondary forest.

Size: male average height 5.6ft (170cm), occasionally up to 5.9ft (180cm); weight 310–400lb (140–180kg). Female height up to 5ft (150cm); weight 200lb (90kg). Often very obese in captivity—one male weight of 750lb (340kg) recorded.

Coat: black to brown-gray, turning gray with age; males with broad silvery-white saddle. Hair short on back, long elsewhere. Skin jet black almost from birth.

Gestation: 250–270 days.

Longevity: about 35 years in wild, 50 years in captivity.

Three races:
Western lowland gorilla (*G. g. gorilla*). Cameroon, Central African Republic, Gabon, Congo, Equatorial Guinea. Coat brown-gray, and male's silvery-white saddle extends to the rump and thighs.

Eastern lowland gorilla (*G. g. graueri*). E Zaire. Coat black, with male's saddle restricted to the back. Jaws and teeth larger, face longer, and body and chest broader and sleeker than Western lowland gorilla.

Mountain gorilla [E] (*G. g. beringei*). Zaire, Rwanda, Uganda at altitudes of about 5,450–12,500ft (1,650–3,790m). Similar in coat and form to Eastern lowland gorilla, but with longer hair, especially on arms. Jaws and teeth even longer, but arms shorter.

[E] Endangered. [V] Vulnerable.

The gorilla is the largest of living primates and, along with the two species of chimpanzee, the ape most closely related to man. Indeed, fossils and biochemical data indicate that the chimpanzees and gorilla are more closely related to man than they are to the orang-utan, the fourth of the "great apes," and are probably the most intelligent land animals on earth, apart from humans, at least as judged by human standards. They can learn hundreds of "words" in deaf-and-dumb sign language and even string some together into simple grammatical two-word "phrases." Nevertheless, the gorilla's formidable appearance, great strength and chest-beating display have given it an otherwise unfounded reputation for untameable ferocity. Field studies show that wild gorillas are no more savage than any other wild animal. In Rwanda thousands of tourists every year approach on foot to within a few yards of totally unrestrained wild gorillas, and not one of these visitors has ever been hurt. Threatened by man alone, adult males are dangerously aggressive only in defense of their breeding rights and family groups.

There are three races of gorilla found in two widely separated areas of Africa. Although obviously connected in the past, the western and eastern populations are kept apart by the River Zaïre–gorillas do not readily swim—and the nature of the intervening terrain. This is primary forest where, because it is so dark beneath the high, closed canopy, very little ground vegetation grows—certainly not enough to support the predominantly terrestrial gorilla.

Gorillas are mainly terrestrial and quadrupedal—they walk on the soles of their hind limbs, but pivot on the knuckles of their forelimbs. However some individuals, particularly the young, spend much time in trees and occasionally adults make foraging trips into the forest canopy. Lightweight individuals can be seen swinging from tree to tree by their arms (brachiation).

► **Coat colour varies** between subspecies of gorilla. In the Western lowland gorilla ABOVE the hair has a brownish-gray tinge whereas in the Eastern lowland gorilla RIGHT it is a more uniform black.

▲ **Most endangered** of the gorilla subspecies is the Mountain gorilla. Only a few hundred survive in the mountains of East Africa.

▶ **Kasimir the silverback** OVERLEAF rests his huge bulk in dense undergrowth of secondary forest in Kahusi-Beiga National Park, eastern Zaire. Some of the best studied gorilla groups are to be found in this park.

The gorilla differs from its close relative the chimpanzee in being very much larger and in the different proportions of its body (longer arms, shorter and broader hands and feet) and a different color pattern. The build is much heavier, and many of the proportional differences are connected with this. In particular, the much larger teeth (especially the molars) needed to sustain a huge hulk must in turn be worked by much bigger jaw muscles, especially the temporal muscles which in male gorillas meet in the midline of the skull, where they are attached to a tall bony crest—the sagittal crest. A small sagittal crest may occur in female gorillas or in chimpanzees, but a big one, meeting a big shelf of bone (the nuchal crest) at the back of the skull, is a distinctive characteristic of male gorillas and considerably alters the external shape of the head. In addition, male gorillas have canines that in relation to body size are far bigger than those of chimpanzee males and bigger also than females' canines. The gorilla has small ears and the nostrils are bordered by broad, expanded ridges (naval wings) which extend to the upper lip.

Gorillas are predominantly folivorous, feeding mainly on leaves and stems rather than fruits. The species of herbs, shrubs and vines that make up the gorilla's diet grow best in secondary and montane forests where the open canopy allows plenty of light to reach the forest floor. Although found over a small area of Africa, gorilla habitat includes a wide range of altitudes, from sea level in West Africa to 12,500ft (3,790m) in the east.

Of the great apes (family Pongidae), the gorilla shows the most stable grouping patterns. The same adult individuals travel together for months and usually years at a time. It is because gorillas are mainly leaf-eating that they can afford to live in these relatively permanent groupings. Apes, unlike most monkeys, cannot digest unripe fruit, and in East Africa ripe fruits, in contrast to leaves, are far too thinly distributed to support a large permanent grouping of frugivorous (fruit-eating) animals even half as big as gorillas. It does appear that fruit-eating limits group size in the gorilla. In West Africa, where fruit forms a far higher proportion of the gorilla's diet than in the east, group sizes are about half those recorded for East Africa. (The predominantly frugivorous chimpanzee and orang-utan are mostly solitary.)

The gorilla's large size and folivorous habits mean that the animals cannot regularly travel long distances and still find time to forage and digest their bulky diet. In fact, because of the abundance of their food supply, under natural conditions they do not need to move very far in any one day to find enough to eat. Although the home ranges of gorilla groups cover 2–11.5sq mi (5–30sq km) depending on the region, the normal rate of travel is only 0.3–0.6mi (0.5–1km) per day. Even with a range as small as 2 sq mi, a circumference of 5mi would have to be patrolled were the area to be defended. Gorillas are too large and travel too slowly to do this and thus there is no territorial defense. Consequently there is considerable overlap of the ranges of neighboring groups and even overlap of the most-used "core areas."

Gorillas never stay long enough at one feeding site to strip it completely: rather, they crop the vegetation leaving enough growth for rapid rejuvenation to occur. They normally feed during the morning and afternoon and rest for a few hours around midday. At night they make "nests"—platforms or cushions of branches and leaves pulled and bent under them, that keep them off the cold ground, prevent them sliding down a steep hill slope, or support them in a tree for the night. At higher altitudes in East Africa, gorillas defecate in their nests during the night, perhaps because it is too cold to leave them. However, the lack of fruit in the diet means that the dung is dry and does not foul their coats. By contrast, in West Africa, where the diet includes fruit, nests with dung in them are extremely rare.

Gorillas do not have a distinct breeding season. Births are usually single (as in the chimpanzees and orang-utan). In the very rare cases of twins, they are usually so small when born, and the mother, who has to carry the infants for the first few months of life, finds it so hard to care for two, that at least one always dies. Newborn infants weigh 4–5lb (1.8–2.3kg) and their grayish

pink skin is sparsely covered with fur. They begin to crawl in about nine weeks and can walk from 30–40 weeks. Gorillas are weaned at $2\frac{1}{2}$–3 years of age, and females give birth at about four-year intervals. However a 40 percent mortality rate in the first three years means that a surviving offspring is produced only about once every 6–8 years in the breeding life of a female.

Female gorillas mature sexually at 7–8 years of age but usually do not start to breed until they are 10 or so. Males mature a little later, but because of competition among them for mates, very few will start to breed before 15–20 years of age.

The size of gorilla groups varies from two to about 35 animals but usually numbers five to 10. An average group in the east (Rwanda, Uganda and eastern Zaire) contains about three adult females, four or five offspring of widely different ages, and one fully adult male—called the "silverback" because of its silvery white saddle. Groups in West Africa average about five animals. Because groups effectively always contain

a silverback male, the difference is the number of females and offspring. But the main contrast between the regions is seen not in average, but maximum group size. In the west, groups of over 10 animals are rare, but in the east 15–20 members is not uncommon and some groups of over 30 animals have been recorded in Zaire.

The silverback is the most important individual for the cohesiveness of the group. Unlike the females of most other social mammals, female gorillas leave the natal group at puberty to join other troops. Having mostly originated from different groups, the adult females in any one silverback's harem are therefore mostly unrelated, the social ties between them are weak, and little difference of social status among the females is noticeable. In contrast to many other primates it is bonds between each female and the silverback, rather than bonds among females, that hold the group together. The attractiveness of the silverback to the females is most apparent during the midday rest period (see LEFT), when animals play (primarily youngsters), sleep or groom one another. Mutual grooming, which keeps fur free from dirt and parasites and is an expression of affinity, is not as frequent in gorillas as in other social primates. When it does occur, it is usually between mother and offspring, adult female and silverback, and sometimes immature and silverback, especially if the youngster is an orphan. Social grooming among adult females is rare. Most young adult males leave the group and travel alone—sometimes for years—until they acquire females from other groups and establish their own harem. These young silverbacks leave of their own accord rather than being driven out by the leading male. Aggression in gorillas is extremely rare and serious fights occur only when a group leader meets either another group leader or, more usually, a lone silverback. Then, the two males perform elaborate acts of threat—the famous chest-beating display may be accompanied by hoots, barks and roars, tearing of vegetation and sideways dashes—all designed to intimidate the rival male and possibly also to impress and attract some of his nubile females.

In general, females leaving their natal group seem to prefer to join lone males or small groups, rather than large, established groups. Lone males appear to be prepared to work harder for females than do leaders of established groups, and therefore pose more of a threat to "resident" leaders.

When females leave their natal group, they generally do not stay with the first male to whom they transfer. Many factors probably influence their choice. One could be the quality of the habitat in the male's range, but another is almost certainly the male's prowess in fights, which provides some indication to a female of the male's ability to protect her and her offspring against predators and other males. Protection against other males is important: about a quarter of infant deaths are due to infanticide by a male that is not the infant's father. The most likely explanation for this is that an intruder male that kills a female's young offspring

▲ **Midday rest for a gorilla group.** As if drawn by a magnet, most individuals gather round the silverback LEFT. Females with very young infants FAR LEFT tend to be closest, with the result that infants rapidly become used to the silverback's presence, and he becomes a focus for them as well – even at play ABOVE FOREGROUND infants keep within his sphere of protection. Females without infants remain in the background on the edge of the gathering TOP, while subadult males RIGHT are tolerated until they leave on reaching maturity to lead solitary lives before setting up their own harems.

◄ **Silverback male** is the only fully adult male in the group and may be twice the weight of the adult females who with their offspring make up the rest of the group.

► **Mountain gorilla nursing her young.** Births are $3\frac{1}{2}$–$4\frac{1}{2}$ years apart and infant mortality reduces the female's successful raising of young to about one every seven years.

can mate with her, and so begin to reproduce, sooner than a male that does not practice such infanticide.

It appears that once a male has established a successfully breeding harem, he stays with it for life. With some males in near-permanent possession of females, while others have none, competition among males for females is intense: the severity of fights between males bears witness to this. Clearly, to some extent, large size is an advantage in these fights, and in the displays that precede them. Inter-male competition is thus almost surely one explanation for the sexual dimorphism (difference of body form between the sexes) in body size, canine size and jaw musculature shown by the gorilla. In this, the gorilla matches most other polygamous mammal species. Given the competition between males and the necessity for the male of this slow-moving terrestrial species to defend females and offspring, it is very probable that natural selection favors the survival of males with large canines: certainly the gorilla is one of the most sexually dimorphic of all primate species, the males being almost twice as large as the females and, moreover, differently colored.

The number of gorillas surviving in the wild today is not accurately known—far too little of their range has been surveyed. We have to make informed guesses, and the best available estimate (1980) has at least 9,000 in west-central Africa and 4,000 in the east, of which only about 365 are Mountain gorillas. Most of these are in danger; they and their habitat are disappearing and will continue to do so at an ever-increasing rate. Gabon, with three-quarters of its land surface still covered by forest and its very low, only slowly increasing, human population, contains about half the Western lowland gorilla population and Zaire almost all the Eastern one.

Throughout the gorilla's range in Africa the forests on which it depends are being cut down for timber and to make way for agricultural and, in some cases, industrial development. Formerly, deforestation did not matter because the human population was at a low enough density to practice shifting agriculture, and the abandoned fields with their regenerating secondary-growth forest provided abundant food for the gorilla. As time passes, however, more clearings are becoming permanent. Twenty-five years ago, gorillas lived in Nigeria. Now they are almost certainly extinct there and cattle ranches cover what used to be gorilla habitat.

▲ **This orphaned gorilla "Julie"** was returned to the wild by Adrien Deschryver in 1974. Sadly, the adoptive gorilla group contained no lactating females and she died soon afterwards.

Another threat is hunting, although it poses a minor problem compared to deforestation. Except in Uganda and Rwanda, gorillas are killed for food and because they raid crops. In West Africa gorillas are considered a crop pest throughout much of their range and firearms are common. Gorilla meat is consumed not only by the rural people but in Gabon, for example, it is served in the restaurants of main towns.

Zoos have also taken their toll. By the end of 1976, of 497 gorillas reported to be alive in captivity, 402 were caught in the wild; at

▼ **Less of a threat** to gorilla's survival than deforestation, hunting for sale of skulls and skins, and capture for export to zoos, are nonetheless enough in themselves to bring the gorilla to the edge of extinction, unless more protective steps are taken.

▲ **Mountain gorilla** in its natural habitat. National Parks and conservation areas need further protection themselves to ensure gorillas' future.

least one-third and probably more than half of these came from Cameroon. Since for every gorilla reaching a zoo at least two have died on the way, this figure represents at least 1,200 taken from the wild. Although the trade has decreased markedly in recent years, the current price paid for a young gorilla is enough to keep going the trade in animals for zoos.

Legislation to control hunting and capture of gorillas exists in all eight countries with wild gorilla populations, but in only one—Zaire—is the species totally protected by law and in none are the laws adequately enforced. At the time of writing, infant gorillas for sale are still coming out of Zaire. While national parks or reserves that contain gorillas exist in all countries except the Central African Republic, only about one-third of the roughly 11,400sq mi (29,500sq km) of these conservation areas is suitable gorilla habitat. Less than 5,000, and probably only about 3,000 gorillas, actually live in national parks or reserves. Even in them, the gorilla cannot be considered safe. The trade in gorilla skulls (sold to tourists as souvenirs) from the Virunga Volcanoes National Parks of Rwanda and Zaire, and the presence of villages and logging yards within Gabon's Okanda National Park, illustrate this. Nor can the continued existence of the conservation areas themselves be guaranteed: in 1968, almost one-fifth of the Virunga Volcanoes National Park, one of only two refuges for the rare Mountain gorilla, was appropriated for pyrethrum cultivation.

In all countries harboring gorillas, it is extremely difficult for governments to fund major conservation programs given their other urgent priorities. Long-term ecological stability is usually sacrificed for short-term economic gain, often with the active encouragement of international development organizations. If the gorilla is to be saved, international conservation agencies must support existing reserves and fund extensive surveys in West Africa and Zaire to pinpoint the most important areas of gorilla concentration which might then be turned into reserves. Ultimately, the local people must come to value gorillas and the forests in which they live. For this reason, the funding of programs for conservation, education and the development of tourism (whereby live wild gorillas become a source of income) are particularly important. Such programs have recently been established successfully in Rwanda.

All this, however, is a rather desperate rearguard action. In the long run, the well-being both of the gorilla and of its human neighbors must depend on curbing human population growth and increasing the productivity of agricultural land now available. Only these will check the increasing need for yet more land and the consequent destruction of the gorilla's habitat.

The story of man's relationship with gorillas is shrouded in myth and legend (and horror movie: *King Kong*, 1933). It was in 1959 that the Americans George and Kay Schaller began the first thorough investigation of gorillas by observing them from close quarters over a two-year period. This work set a pattern for future studies of these peaceable creatures and effectively buried the gorilla myth for good. AHH

TREE SHREWS

ORDER: SCANDENTIA

One family (Tupaiidae); 6 genera; 18 species.
Distribution: W India to Mindanao in the Philippines, S China to Java, including most islands in Malayan Archipelago. Habitat: tropical rain forest.

Subfamily Tupaiinae

Genus *Tupaia*: 11 species including **Pygmy tree shrew** (*T. minor*), **Belanger's tree shrew** (*T. belangeri*) and **Common tree shrew** (*T. glis*).

Genus *Anathana*: 1 species, the **Indian tree shrew** (*A. ellioti*).

Genus *Urogale*: 1 species, the **Philippine tree shrew** (*U. everetti*).

Genus *Dendrogale*: 2 species, the **smooth-tailed tree shrews.**

Genus *Lyonogale*: 2 species, including the **Terrestrial tree shrew** (*L. tana*).

Subfamily Ptilocercinae

Genus *Ptilocercus*: 1 species, the **Pen-tailed tree shrew** (*P. lowii*).

TREE shrews are small, squirrel-like mammals found in the tropical rain forests of southern and southeastern Asia. The first reference to them is in an illustrated account in 1780 by William Ellis, a surgeon who accompanied Captain Cook on his exploratory voyage to the Malay Archipelago. It was Ellis who coined the name "tree shrew," a somewhat inappropriate term since these unusual mammals are quite different from the true shrews and, as a group, are not particularly well adapted for life in trees—indeed some tree shrew species are almost completely terrestrial in habit.

Most tree shrew spécies are semiterrestrial, rather like the European squirrels, a resemblance that is emphasized by the fact that both tree shrews and squirrels are covered by the Malay word "tupai," from which the genus name *Tupaia* is derived. At first classified as insectivores, tree shrews were for a time thought to be primates, but recent research distinguishes them from both orders and they are here considered as the only members of the order Scandentia (see p147).

None of the six genera covers the entire geographical range of the order, though the genus *Tupaia* is the most widespread of all. The greatest number of species is to be found on Borneo, where 10 of the 18 recognized species occur. This concentration is partly a consequence of the large size of the island and the resulting wide range of available habitats, but it is also possible that Borneo was the center from which the adaptive radiation of modern tree shrew species began.

Tree shrews are small mammals with an elongated body and a long tail which, except in the Pen-tailed tree shrew, is usually covered with long thick hair. Their fur is dense and soft. They have claws on all fingers and toes; the first digits diverge slightly from the others. Their snouts range from short to elongated. The ears have a membranous external flap, which varies in size from species to species, and is usually covered with hair. In the Pen-tailed tree shrew the ear flaps are bare and larger than in any other tree shrews, doubtless because this nocturnal species relies more heavily on

► **Representative species of tree shrews.** (1) The largely arboreal Pygmy tree shrew (*Tupaia minor*) "sledging" along a branch and leaving a scent trail from its abdominal gland. (2) The arboreal Pen-tailed tree shrew (*Ptilocercus lowii*) holding a captured insect in both hands while devouring it. This is the only nocturnal species and the only living representative of the subfamily Ptilocercinae. (3) The semi-terrestrial Common tree shrew (*Tupaia glis*) "chinning" to leave a scent trail from the sternal gland on its chest. (4) The mainly arboreal Northern smooth-tailed tree shrew (*Dendrogale murina*) snatching an insect from the air with both hands. (5) The Terrestrial tree shrew (*Lyonogale tana*) is a large-bodied species that finds most of its food in litter on the forest floor, rooting with its snout and turning over objects such as stones. (6) The Philippine tree shrew (*Urogale everetti*) is the largest species and it too spends most of its time at ground level rooting through debris.

Skulls of Tree Shrews

The skull of a tree shrew is that of a strictly quadrupedal mammal. The foramen magnum (the opening through which the spinal cord passes) is directed backward, whereas in the more upright living primates it is directed more or less downward. The skull is longest in terrestrial species which root in leaf-litter. Arboreal species have shorter snouts, larger brains and more forward facing eyes that give binocular vision. Compared here are the terrestrial Philippine and Terrestrial tree shrews (**a**, **b**), the semi-arboreal Common tree shrew (**c**) and the tree-dwelling Pygmy tree shrew (**d**). Common skull features include the post-orbital bar and an auditory bulla.

The dental formula is I2/3, C1/1, P3/3, M3/3 = 38. The canine teeth are relatively poorly developed compared with most mammals; the primitive sharp-cusped molars reflect an insectivorous diet and the forward-projecting lower incisors are used in grooming.

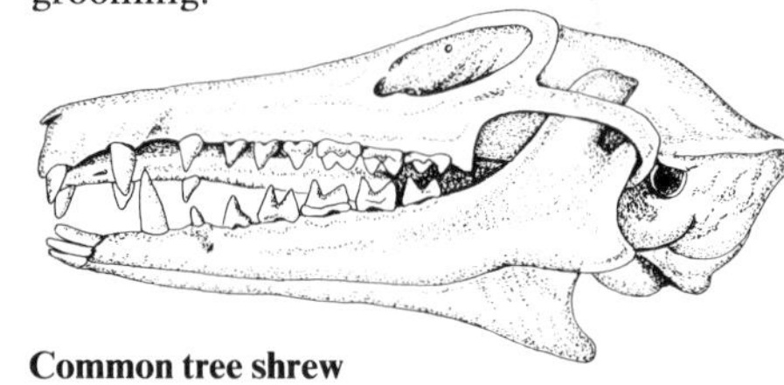

Common tree shrew

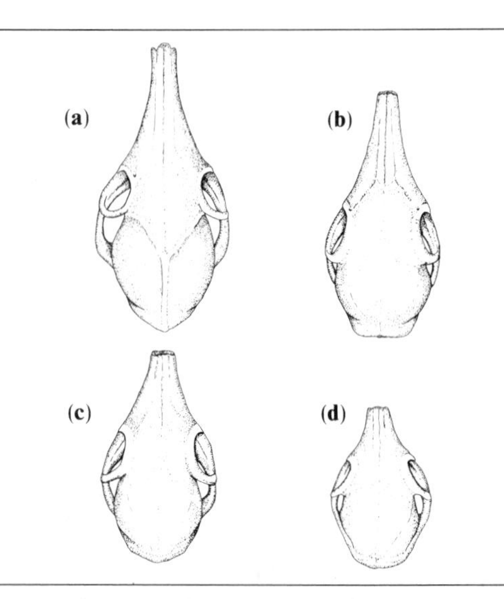

ORDER SCANDENTIA

its hearing to find insect prey and avoid predators at night. Species that are mainly arboreal, such as the Pygmy tree shrew, are small, have short snouts, more forward facing eyes, poorly developed claws, and tails longer than the combined head and body length. Terrestrial species, such as the Terrestrial tree shrew, are large, have elongated snouts, well-developed claws for rooting after insects, and tails shorter than head-and-body length. Their terrestrial habits permit greater body size with less need for a long tail to balance in the trees.

In most anatomical features, tree shrews show little obvious specialization, though there are some unusual features in the skull and dentition (see above). Except for the Pen-tailed tree shrew, tree shrews generally have laterally placed eyes, which are large and give good vision relative to body size. There is a well-developed subtongue (sublingua) beneath the tongue.

Like European squirrels, most tree shrews spend more time foraging on the ground than in trees. They scurry up and down tree trunks and across the forest floor with characteristic jerky movements of their tails, feeding on a wide variety of small animal prey (especially arthropods) and on fruits, seeds and other plant material. All except the most arboreal species spend a great deal of time rooting in leaf litter with the snout and hands. Tree shrews typically prefer to catch food with their snouts and only use their hands when food cannot be reached otherwise. However, flying insects may be caught with a rapid snatch of the hand and all tree shrew species hold food between their front paws when eating. The larger tree shrews probably eat small vertebrates in the wild, for example small mammals and lizards, since in captivity they have been seen to overpower adult mice and young rats and to kill them with a single bite to the neck.

Most tree shrews (like squirrels) are exceptions to the general rule that small mammals are nocturnal. The Pen-tailed tree shrew, however, is exclusively nocturnal, and many of its distinctive features (larger eyes and ears, large whiskers, gray/black coloration) can be attributed to this difference. It has been suggested that the Smooth-tailed tree shrews might be intermediate in exhibiting a crepuscular pattern, with peaks of activity at dawn and dusk.

Most tree shrew species nest in tree hollows lined with dried leaves. The gestation period is 45–50 days, according to species, and the 1–3 young are born without fur, with closed ears and eyes. The ears open within 10 days and the eyes in 20 days.

Tree shrews are unusual among placental mammals for the extremely rudimentary nature of their parental care. Laboratory studies have shown that in at least three species (the Pygmy, Belanger's and Terrestrial tree shrews) the mother gives birth to her offspring in a separate nest which she visits to suckle them only once every two days (early attempts to breed tree shrews in captivity failed largely because only one nest was provided). The visits are very brief (5–10 minutes), and in this short space of time she provides each infant with 0.18–0.53oz (5–15g) of milk to provision it for the following 48 hours. The milk contains a large amount of protein (10 percent), which permits the young to grow rapidly, and an unusually high fat concentration (26 percent) which enables them to maintain their body temperature in the region of 98.6°F (37°C) despite the absence of the mother from the nest. However, the infants are relatively immobile in the nest, so the milk contains only a small proportion of carbohydrate (2 percent) for immediate energy needs.

In all three shrew species so far studied, the infants have been found to stay in the nest for about a month, after which they emerge as small replicas of the adults. The young continue to grow rapidly and sexual maturity may be reached by the age of four months. Between the birth of her offspring and their eventual emergence from the nest, the mother spends a total of only one and a half hours with the infants, and her brief suckling visits are accompanied by no toilet care. Indeed, maternal care in tree shrews is so limited that if an infant tree shrew is removed from its nest and placed just beside it, the mother will completely ignore it. She only recognizes her offspring in the nest because of a scent mark which she deposits

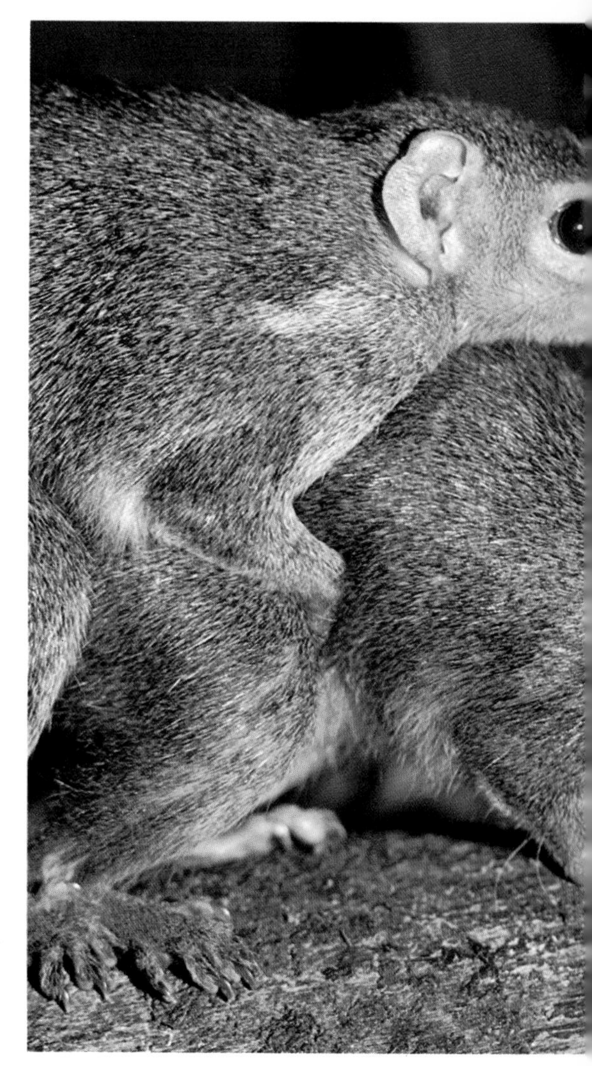

▲ **Indian tree shrews mating.** The gestation period is about 45 days, which is longer than that of insectivores of similar size, but much shorter than for comparable primates. Tree shrews therefore have a more rapid reproductive turnover than primates.

► **Rudimentary parental care** of tree shrews is unusual among placental mammals. The mother only suckles her young every 48 hours, leaving them completely alone in the nest between feeds. Because of this widely spaced feeding the young have to take in large quantities of milk at each session and consequently become extremely bloated, as can be seen from this litter.

◄ **Diet of tree shrews** mainly comprises insects and arthropods, but most species will feed on any available small invertebrate prey, such as the earthworms shown here or small mammals such as rats and mice.

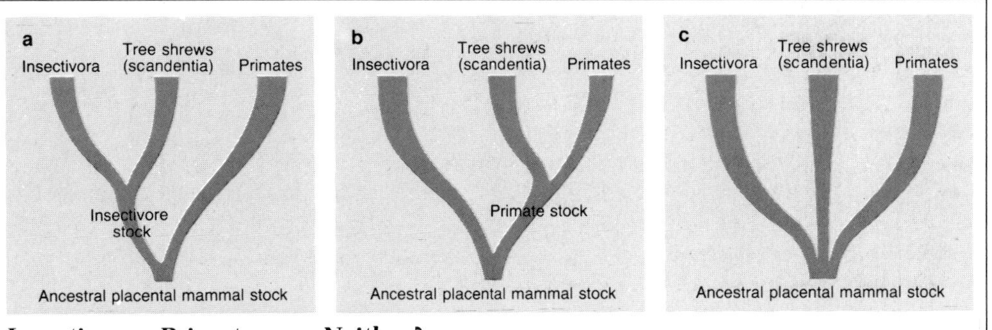

Insectivores, Primates—or Neither?

Misleadingly named, and originally placed within the order Insectivora along with the tree shrews, tree shrews would probably have remained in obscurity but for the suggestion made by the anatomist Wilfred Le Gros Clark in the 1920s that these relatively primitive mammals might be related to the primates. Comparing the structure of the skull, brain, musculature and reproductive systems, he concluded that the tree shrews should be regarded as the first offshoot from the ancestral primate stock. This interpretation was accepted by George Gaylord Simpson, who included the tree shrews in the order Primates in his influential classification of the mammals, published in 1945.

Thereafter, numerous studies of tree shrews were conducted in the hope of clarifying the evolutionary history of the primates (including man), and the tree shrews became widely regarded as present-day survivors of our primate ancestors. Today's consensus is that tree shrews are not specifically related to either primates or insectivores, but represent a quite separate lineage in the evolution of the placental mammals.

This view has been reached by several means. The first objection to the "primate" interpretation is that tree shrews may have come to resemble primates through entirely separate (convergent) evolution of certain features because of similar functional requirements. For instance, primates are typically arboreal while insectivores (eg shrews, hedgehogs, moles, tenrecs etc) are typically terrestrial, so it is possible that tree shrew/primate similarities have evolved through convergent adaptations for arboreal life. Various apparently primate-like features—relative shortening of the snout, forward rotation of the eye-sockets and greater development of the central nervous system associated with large eyes—are largely confined to the most arboreal of the tree shrew species, such as the Pygmy tree shrew. So these special characters cannot reliably be regarded as vestiges from a common ancestral stock of tree shrews and primates, particularly since the ancestral tree shrew was probably closest to the modern semi-terrestrial species.

The second, and most important, objection is based on drawing a distinction between similarities that are shared because they derive from a specific ancestral stock and similarities shared merely because of retention of characters from the more ancient ancestral stock of placental mammals. This second kind of similarity does not itself indicate any specific relationship between tree shrews and primates. A particularly good example is the presence of the cecum, a blind sac in the digestive tract at the junction of the small and large intestines, housing bacteria which assist in the breakdown of plant food. Tree shrews and primates typically possess a cecum, whereas it is absent from most insectivores. However, it has recently been shown that the cecum is widespread among mammals, both placentals and marsupials, and is even present in reptiles. It therefore seems likely that the cecum was already present in the earliest placental mammals, so its retention provides no evidence whatsoever for a specific ancestral connection between tree shrews and primates.

It has also emerged that in some respects tree shrews are very different from primates, particularly in reproductive characters, since the development of the placenta in tree shrews is quite unlike that in any primate species and the offspring are born in a naked, helpless condition that contrasts markedly with the advanced condition of newborn primates. In tree shrews, parental care is very rudimentary and also far removed from the elaborate parental care of the primates.

A major difficulty in reconstructing the evolution of tree shrews is the lack of fossil evidence, a gap that has now been filled to some extent by the discovery in Siwalik deposits of the Indian Miocene of tree shrew fossils (*Palaeotupaia sivalensis*) dating back some 10 million years. These support the interpretation that tree shrews derive from a semi-terrestrial ancestral form with moderate development of the snout.

It is probably best to regard the tree shrews as an entirely separate order of mammals which branched off very early during the radiation of placental mammal types. This view is also supported by biochemical evidence recently derived from the immunological cross-reactions of proteins from tree shrews, insectivores and primates, and from comparison of amino acid sequences in proteins from these species.

It now seems that rather than being survivors from early primate stock, the tree shrews may well be closer to the common ancestors of placental mammals in general.

THE 18 SPECIES OF TREE SHREWS

Abbreviations: HBL = head-and-body length; TL = tail length; wt = weight.

Tree Shrew Classification

In the classification of tree shrews particular emphasis is placed on coat coloration patterns, tail length and shape, form of the ears, development of the snout and claws, and the number of teats in females (the number of pairs of teats being the same as the number of offspring typical for the species).

Tupaia contains the least specialized tree shrews, distinguished primarily by the absence of special features. The wide geographical distribution of the genus is associated with considerable speciation and at least 11 distinct species are recognized.

The single species of the genus *Anathana* occurs in India south of the River Ganges, which divides it from *T. belangeri* to the east from which it differs in having relatively larger ear flaps and more complex molar tooth cusp patterns.

The two predominantly ground-living species now placed in the genus *Lyonogale* have elongated snouts and robust claws on the forefeet.

Urogale is quite distinct in many ways. The single known species is the largest of living tree shrew species and the sole representative on the island of Mindanao. The dentition is striking in that the second pair of incisors in the upper jaw are prominent and canine-like (completely dwarfing the actual canines), while in the lower jaw the third pair of incisors show great reduction.

The two species of the genus *Dendrogale* are distinguished by their fine tail fur, diminutive body size and characteristic facial markings.

The Pen-tailed tree shrew (*Ptilocercus lowii*) is placed in a separate subfamily. It is the only nocturnal tree shrew and appears to be almost exclusively arboreal. The eyes are relatively forward facing, giving marked binocular overlap. The hands and feet are relatively large and in the upper jaw the anterior incisors are enlarged in a distinctive fashion. Pen-tailed tree shrews lack a shoulder stripe, in contrast to all except *Dendrogale* species.

The Smooth-tailed tree shrews are in many ways intermediate between the other tree shrews and *Ptilocercus*, for example in the sparseness of the hair on their tails and in reportedly showing a crepuscular pattern of activity.

Subfamily Tupaiinae

Active in daytime (*Dendrogale* perhaps at twilight). Eyes laterally placed. Five genera.

Genus *Tupaia*

S Malay Peninsula, Indonesia, Philippines, Indochina. Semi-terrestrial or arboreal; medium-sized (6oz) or small (1.5oz). Conspicuous cream or buff shoulder stripes always present; snout short (arboreal forms) or slightly elongated (semi-terrestrial forms); canine teeth moderately well developed. Ear flaps small. Females with 1, 2 or 3 pairs of teats.

Belanger's tree shrew
Tupaia belangeri

Indochina. Semi-terrestrial. Tail equal to length of head and body combined. wt 6oz. Coat: ranging from olivaceous to very dark brown above, and from creamy-white or orange-red below. Females with 3 pairs of teats.

Common tree shrew
Tupaia glis

S Malay Peninsula, Sumatra and surrounding islands. Habit, weight, coat and tail features as *T. belangeri*. Females with 2 pairs of teats.

Long-footed tree shrew
Tupaia longipes

Borneo. Habit, weight, coat and tail features as *T. belangeri*. Females with 3 pairs of teats.

Montane tree shrew
Tupaia montana
Montane or Mountain tree shrew.

N Borneo (mountains). Habit, weight, coat and tail features as *T. belangeri*. Females with 2 pairs of teats.

Nicobar tree shrew
Tupaia nicobarica

Nicobar Islands. Semi-terrestrial to arboreal. Tail longer than length of head and body combined. Weight and coat as *T. belangeri*. Females with 1 pair of teats.

Painted tree shrew
Tupaia picta

N Borneo (lowlands). Habit, weight, coat and tail features as *T. belangeri*, except that a dark stripe runs length of back. Females with 2 pairs of teats.

Palawan tree shrew
Tupaia palawanensis

Philippines. Habit, weight, coat and tail features as *T. belangeri*. Females with 2 pairs of teats.

Rufous-tailed tree shrew
Tupaia splendidula

SW Borneo, NE Sumatra. Habit weight, coat and tail features as *T. belangeri*. Females with 2 pairs of teats.

Pygmy tree shrew
Tupaia minor

Borneo, Sumatra, S Malay Peninsula and surrounding islands. Arboreal TL greater than HBL. wt 1.5oz. Coat: olivaceous above, off-white below. Females with 2 pairs of teats.

Indonesian tree shrew
Tupaia javanica
Indonesian or Javan tree shrew.

Java, Sumatra. Habit, weight, coat, tail and teats as *T. minor*.

Slender tree shrew
Tupaia gracilis

N Borneo and surrounding islands. Habit, weight, coat, tail and teats as *T. minor*.

Genus *Anathana*

Semi-terrestrial. Medium sized (6oz). Coat: brown or gray-brown above, buff below; shoulder stripe light buff or white; pale markings around eyes. Tail equal in length to head and body combined. Snout short. Canine teeth poorly developed. Ear flaps well developed. Females with 3 pairs of teats.

Indian tree shrew
Anathana ellioti

India south of the Ganges.

Genus *Urogale*

Terrestrial. Large (12oz). Coat: dark brown above, yellowish or rufous below; shoulder stripe pale. Tail much shorter than length of head and body combined, and covered in closely set rufous hairs. Snout elongated. Second pair of incisors enlarged. Females with 2 pairs of teats.

Philippine tree shrew
Urogale everetti

Mindanao.

Genus *Dendrogale*

Arboreal. Small (2oz). No shoulder stripes present. Tail slightly longer than length of head and body combined, covered with fine smooth hair. Snout short. Ear flaps large. Females with 1 pair of teats.

Southern smooth-tailed tree shrew
Dendrogale melanura
Southern or Bornean smooth-tailed tree shrew.

N Borneo. Coat: dark brown above, pale buff below: facial streaks inconspicuous, but orange-brown eye rings prominent. Claws sharp.

Northern smooth-tailed tree shrew
Dendrogale murina

S Vietnam, Cambodia, S Thailand. Coat: light brown above, pale buff below; dark streak on each side of face running from snout to ear and highlighted by paler fur above and below. Claws small and blunt.

Genus *Lyonogale*

Terrestrial. Large (10oz). Conspicuous black stripe along back; shoulder stripe pale. Tail bushy and shorter than length of head and body combined. Snout elongated. Canine teeth well developed. Claws robust. Females with 2 pairs of teats.

Terrestrial tree shrew
Lyonogale tana

Borneo, Sumatra and surrounding islands. Coat: dark red-brown above, orange-red or rusty-red below; front of dorsal stripe highlighted by pale areas either side; shoulder stripe yellowish. Claws robust and elongated.

Striped tree shrew
Lyonogale dorsalis

NW Borneo. Coat: dull brown above, pale buff below; shoulder stripe creamy buff or whitish. Claws less robust and shorter than in *L. tana*.

Subfamily Ptilocercinae

Nocturnal. Eyes forward-facing, giving binocular vision. One genus.

Genus *Ptilocercus*

Arboreal. Small (2oz). Coat: dark gray above, pale gray or buff below; dark facial stripes running from snout to behind eye; no shoulder stripe present. Tail considerably longer than combined length of head and body; covered for entire length with scales, except for tuft of hairs at tip. Snout short. Upper incisors enlarged. Ear flaps large, membranous and mobile. Females with 2 pairs of teats.

Pen-tailed tree shrew
Ptilocercus lowii

S Malay Peninsula, NW Borneo, N Sumatra and surrounding islands.

▲ **Insect prey in mouth,** the Common tree shrew of Southeast Asia. This semi-terrestrial species has a snout of medium length. Exclusively tree-dwelling species are short-nosed, terrestrial species long-nosed for rooting in leaf-litter.

on them with her sternal gland; if the scent is wiped off she will devour her own infants!

The number of pairs of teats is characteristic of each species and is directly linked to the typical number of infants (one, two or three pairs of teats corresponding to one, two or three offspring).

Tree shrews tend to breed over a large part of the year, though a definite seasonal peak of births has been reported in some cases. The short gestation period and rapid maturation of the offspring mean that tree shrews can breed rapidly if the conditions are right, and they are able to colonize new areas quite quickly.

Of all the tree shrews the Common tree shrew has been most closely observed in the wild. This species forms loose social groups typically composed of an adult pair and their apparent offspring. The members of each group occupy all or part of a common home range covering approximately 2.5 acres (1 hectare), but they usually move around independently during the daytime, predominantly on or near the forest floor. The home range of each group seems to be defended as a territory, since there is very little overlap between adjacent home ranges and since fights have been observed on their boundaries. Tree shrews engage in extensive scent-marking behavior. The details vary from species to species, but in all cases scent marking involves special scent glands, urine and perhaps even feces. Belanger's tree shrew possesses two glandular areas on the ventral surface of the body: the sternal gland is used in "chinning," the tree shrew standing with stiffened legs and rubbing the gland over the object to be marked, which may be a branch or another tree shrew; the abdominal gland is used in "sledging," in which the tree shrew slides down a branch while pressing its abdomen against the surface. Tree shrews also scent mark by depositing droplets of urine while walking along branches, and the Terrestrial tree shrew has been reported to perform a kind of dance in which the hands and feet are impregnated with urine previously deposited on a flat surface. In captivity, at least, the products of these various scent-marking activities accumulate to form an orange-yellow crust with a fatty consistency and an extremely pungent smell. Captive tree shrews also deposit their feces in a few specific places in the cage, suggesting the droppings may play a role in territorial demarcation in the wild.

Tree shrews have a rather limited range of calls. All species, when surprised in the nest or during attacks on other tree shrews, produce a hoarse, snarling hiss with the mouth held wide open. Infants produce a similar sound when disturbed in the nest. A variety of squeaks and squeals is produced during fights, culminating in really piercing squeals when one combatant is beaten. *Tupaia* species also produce a continued chattering call when mildly alarmed, and there is some evidence that this acts as a mobbing call announcing the presence of potential predators.

Tree shrews are relatively inconspicuous mammals and their contacts with man are restricted. They may be pests in fruit plantations, and they occasionally occur in and around human habitations, but they do not seem to occupy any important place in the human economy or in mythology. Because of their high breeding potential, they can recover rapidly from population decline and quickly colonize new areas, so they are not obviously threatened by man at present. Rare species such as the Pen-tailed tree shrew and the Smooth-tailed tree shrews may be vulnerable in the wild, but firm information is lacking. RDM

FLYING LEMURS

ORDER DERMOPTERA

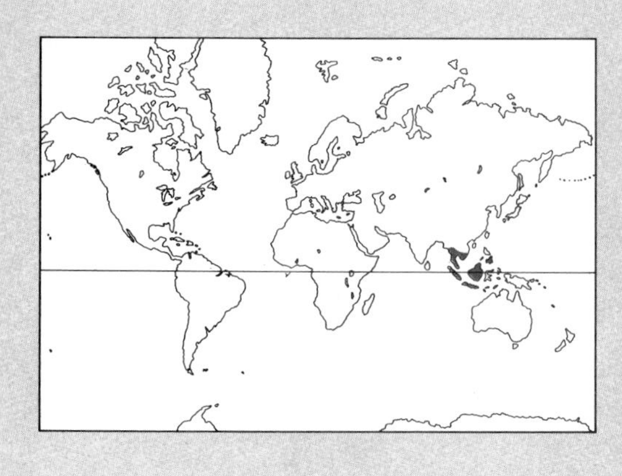

ORDER: DERMOPTERA

One family, the Cynocephalidae.
Two species of the genus *Cynocephalus*.

Malayan colugo

Cynocephalus variegatus
Malayan colugo or Malayan flying lemur.

Distribution: Tenasserim, Thailand, S Indochina, Malaya, Sumatra, Borneo, Java and adjacent islands.
Habitat: tropical rain forests and rubber plantations.
Size: head-body length 13–16.5in (34–42cm); tail length 8–11in (22–27cm); "wingspan" 28in (70cm); weight 2–4lb (1–1.7kg).
Coat: females slightly larger than males. Upper surface of flight membrane mottled grayish-brown with white spots (an effective camouflage on tree trunks), underparts paler; females tend to be more gray, males more brown or reddish.
Gestation: 60 days.
Longevity: not known.

Philippine colugo

Cynocephalus volans
Philippine colugo, Gliding or Flying lemur.

Distribution: Philippine Islands of Mindanao, Basilan, Samar, Leyte, Bohol.
Habitat: forests in mountainous and lowland regions.
Size: head-body length 13–15in (33–38cm); tail length 8–11in (22–27cm); weight 2–3.5lb (1–1.5kg).
Coat: darker and less spotted than Malayan colugo.
Gestation and longevity: as Malayan colugo.

The order Dermoptera includes only two known families, the extinct Plagiomenidae from the late Paleocene and early Eocene of North America 60–70 million years ago and the modern family Cynocephalidae, the colugos or flying lemurs of Southeast Asia. The famous Singapore naturalist Ivan Polunin always refers to colugos as non-flying, non-lemurs, an apt description, as they do not fly but glide and are not lemurs. In the past dermopterans have also been included with the insectivores and bats, and to confuse matters further colugos were known for many years by the family name Galeopithecidae ("cloaked monkeys"). Such problems in classification arise because the family Cynocephalidae has no fossil record. Today it is recognized that colugos are probably remnants of an ancient specialized mammalian side-branch and, because of their unique appearance and habits, they are placed in a separate order—the Dermoptera ("skin-wing").

The order name refers to the flying lemurs' most distinctive characteristic, the gliding membrane or patagium, which stretches from the side of the animal's neck to the tips of the fingers and toes and continues to the very tip of the tail. No other gliding mammal has such an extensive membrane; in the flying squirrels and marsupial phalangers the patagium stretches only between the limbs. With the patagium outspread, the colugo assumes the shape of a kite and can execute controlled glides of 230ft (70m) or more, with the membrane acting as a parachute. During a measured glide of 450ft (136m) between trees one colugo lost only 39ft (12m) in height.

The flying lemurs' spectacular gliding habits are a remarkable adaptation to the environment in which they live: the multilayered rain forest where trees reach great heights and much of the food is in the canopy. In consequence, many rain forest animals have adopted an arboreal way of life, but this poses the problem of how to get from one tree to the next. Birds and bats fly, squirrels jump and monkeys leap; few animals choose to descend to the ground, cross the gap and climb the next tree, a way of travel that is costly in terms of energy and leaves them vulnerable to a host of ground predators. So the flying lemur has found another solution: it glides.

Although flying lemurs are usually found in primary and secondary forests in both lowland and mountain regions, the Malayan colugo is equally at home in rubber estates and coconut plantations, where it is regarded as a pest because it eats budding coconut flowers. Among the well-spaced trees in a plantation, the colugo's flight can be seen to best advantage.

Flying lemurs are so distinctive in appearance and behavior that they can hardly be mistaken for anything else. They are about the size of a cat and so arboreal in habit that a colugo on the ground is almost helpless. The Philippine colugo is smaller than the Malayan colugo and seems to be more primitive, with less specialized upper incisors and canines. A flying lemur's limbs are of equal length, with strong sharp claws for climbing, and the toes are connected by webs of skin, an extension of the distinctive flight membrane. The head is broad, somewhat like a greyhound's in appearance, with rounded short ears and a blunt muzzle. Flying lemurs have large eyes, as befits a nocturnal animal, and their stereoscopic vision gives them the depth perception necessary for judging accurate landings.

Flying lemurs are herbivores and they have teeth unlike those of any other mammal. Like ruminants, they have a gap

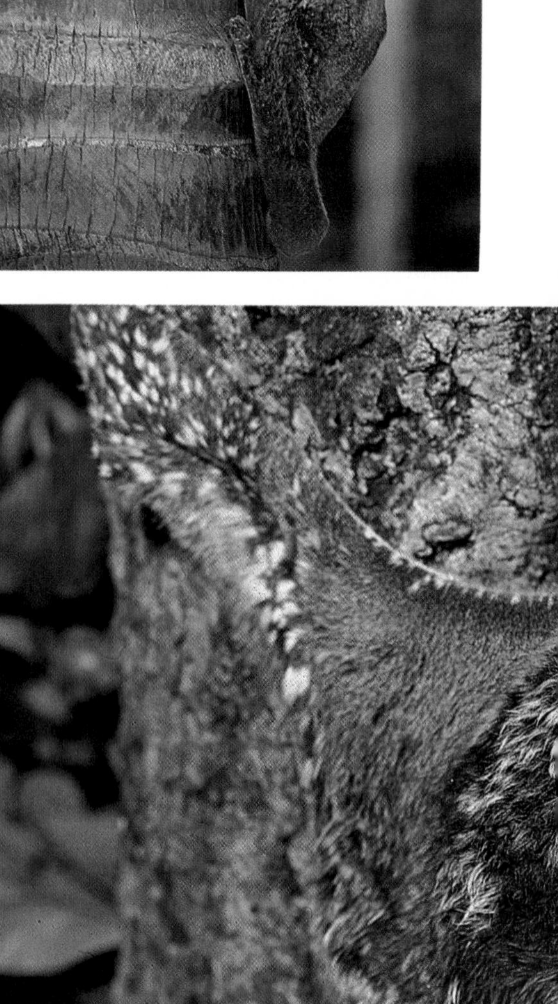

▲ **Kite-like glider** in forests of Southeast Asia, the colugo has a more extensive flying membrane than other gliding mammals.

◀ **Sharp claws** secure the landing of the Malayan colugo on vertical trunks, and enable it to bound upward before launching itself on the next glide.

▼ **Mottled gray upper parts** of the Malayan colugo effectively camouflage the glider on tree trunks.

at the front of the upper jaw with all the upper incisors at the side of the mouth, but the second upper incisor has two roots, a feature unique among mammals. The most interesting aspect of their dentition, however, is the fact that all the incisors are comb-like, with as many as 20 comb tines arising from one root. The function of these unique comb tines is not fully understood; they may be used as scrapers, to strain food or for grooming the animal's fur.

Flying lemurs' diet seems to consist mainly of leaves, both young and mature, shoots, buds and flowers, and perhaps soft fruits, which captive animals will accept reluctantly. When feeding, the colugo pulls a bunch of leaves within reach with its front foot, then picks off the leaves with its strong tongue and lower incisors. The stomach is specialized for ingesting large quantities of leafy vegetation and has an extended pyloric digesting region, the part near the exit to the intestines. The intestines are long and convoluted, with the large intestine longer than the small intestine, the reverse of most mammals. Flying lemurs probably obtain sufficient water by licking wet leaves.

A single young is born at a time, rarely two, after a gestation of 60 days. Lactating females with unweaned young have been found to be pregnant, so it is possible that births may follow in rapid succession. The infant is born in an undeveloped state like a marsupial and until it is weaned it is carried on the belly of the mother, even when she "flies." The patagium can be folded near the tail into a soft warm pouch for carrying the young. When the female rests beneath a bough with her patagium outstretched the infant peers out from an exotic hammock. Young flying lemurs emit duck-like cries and the cry of the adult is said to be similar, although it is rarely used.

Flying lemurs spend the day in holes or hollows of trees or hanging beneath a bough or against a tree trunk with the patagium extended like a cloak. In coconut plantations they curl up in a ball among the palm fronds.

Night falls quickly in the tropics and the colugo usually emerges before dusk and climbs to the top of its tree. It ascends by a series of clumsy bounds, grasping the trunk with its outspread limbs and moving both front feet together then both hind feet. The colugo halts at the top of the tall, straight trunk beneath the tangle of branches at the crown and cranes its head to look down. It does not turn its whole body to do this as a squirrel would. Having chosen its flight path it launches into a long controlled glide to another tree, where it lands low on the trunk, sometimes only 10–13ft (3–4m) above the ground, and climbs slowly upwards, often pausing to rest. The colugo again moves to the top of the trunk (this type of locomotion would be impossible in low-branching temperate woods) and glides off again. The animal may pass quickly from tree to tree, covering considerable distances to reach its feeding trees. Flying lemurs use regular gliding trees and several animals may use the same tree, following each other in rapid succession up the trunk, but choosing different glide paths. Animals move about singly, apart from mothers with young, but several cover the same area and use the same feeding trees. Six independent animals were found in an area of less than 1.2 acres (0.5ha) in a coconut plantation in Java. Flying lemurs forage through the night and return to their sleeping trees at dawn.

Like many other rain-forest species, they are endangered by loss of their habitat to timber-felling or agriculture. KMacK

BIBLIOGRAPHY

The following list of titles indicates key reference works used in the preparation of this volume and those recommended for further reading. The list is divided into two categories: general mammology and those titles relevant to primates.

General

Boyle, C. L. (ed) (1981) *The RSPCA Book of British Mammals*, Collins, London.

Corbet. G. B. and Hill, J. E. (1980) *A World List of Mammalian Species*, British Museum and Cornell University Press, London and Ithaca, N.Y.

Dorst, J. and Dandelot, P. (1972) *Larger Mammals of Africa*, Collins, London.

Grzimek, B. (ed) (1972) *Grzimek's Animal Life Encyclopedia*, vols 10, 11 and 12, Van Nostrand Reinhold, New York.

Hall, E. R. and Kelson, K. R. (1959) *The Mammals of North America*, Ronald Press, New York.

Harrison Matthews, L. (1969) *The Life of Mammals*, vols 1 and 2. Weidenfeld & Nicolson, London.

Honacki, J. H., Kinman, K. E. and Koeppl, J. W. (eds) (1982) *Mammal Species of the World*, Allen Press and Association of Systematics Collections, Lawrence, Kansas.

Kingdon, J. (1971–82) *East African Mammals*, vols I–III, Academic Press, New York.

Morris, D. (1965) *The Mammals*, Hodder & Stoughton, London.

Nowak, R. M. and Paradiso, J. L. (eds) (1983) *Walker's Mammals of the World* (4th edn), 2 vols. Johns Hopkins University Press, Baltimore and London.

Vaughan, T. L. (1972) *Mammalogy*, W. B. Saunders, London and Philadelphia.

Young, J. Z. (1975) *The Life of Mammals: their Anatomy and Physiology*, Oxford University Press, Oxford.

Primates

Altmann, S. A. and Altmann. J. (1970) *Baboon Ecology*. University of Chicago Press, Chicago.

Altmann, J. (1980) *Baboon Mothers and Infants*, Harvard University Press, Cambridge.

Bramblett, C. A. (1976) *Patterns of Primate Behaviour*, Mayfield Publishing Co., Palo Alto.

Chalmers, N. (1979) *Social Behaviour in Primates*, Edward Arnold, London.

Charles-Dominique, P. (1977) *Ecology and Behaviour of Nocturnal Primates: Prosimians of Equatorial West Africa*, Duckworth, London.

Charles-Dominique, P. *et al.* (eds) (1980) *Nocturnal Malagasy Primates: Ecology, Physiology and Behaviour*, Academic Press, New York.

Chivers, D. J. (ed) (1980) *Malayan Forest Primates*, Plenum, New York.

Clutton-Brock, T. H. (ed) (1977) *Primate Ecology*, Academic Press, London.

Clutton-Brock, T. H. and Harvey, P. H. (eds) (1978) *Readings in Sociobiology*, W. H. Freeman, Reading.

Coimbra-Filho, A. F. and Mittermeier, R. A. (1981) *Ecology and Behaviour of Neotropical Primates*, Academia Brasileira de Ciencias, Rio de Janeiro.

Devore, I. (ed) (1965) *Primate Behaviour: Field Studies of Monkeys and Apes*. Holt, Rinehart & Winston, New York.

Doyle, G. A. and Martin, R. D. (eds) (1979) *The Study of Prosimian Behavior*, Academic Press, New York.

Hrdy, S. B. (1977) *The Langurs of Abu: Female and Male Strategies of Reproduction*, Harvard University Press, Cambridge.

Jay, P. C. (1968) *Primates: Studies in Adaptation and Variability*, Holt, Rinehart & Winston, New York.

Jolly, A. (1972) *The Evolution of Primate Behavior*, Macmillan, New York.

Jolly, A. (1966) *Lemur Behavior: A Malagasy Field Study*, University of Chicago Press, Chicago.

Kleiman, D. G. (ed) (1977) *The Biology and Conservation of the Callitrichidae*, Smithsonian Institution Press, Washington.

Kummer, H. (1971) *Primate Societies: Group Techniques of Ecological Adaptation*, Aldine Atherton, Chicago.

van Lawick-Goodall, J. (1971) *In the Shadow of Man*, Collins, London.

Lindburg, D. G. (ed) (1980) *The Macaques: Studies in Ecology, Behavior and Evolution*, Van Nostrand Reinhold, New York.

Martin, R. D., Doyle, G. A. and Walker, A. C. (eds) (1974) *Prosimian Biology*, Duckworth, London.

Michael, R. P. and Crook, J. H. (eds) (1973) *Comparative Ecology and Behaviour of Primates*, Academic Press, London.

Milton, K. (1980) *The Foraging Strategy of Howler Monkeys*, Columbia University Press, New York.

Moynihan, M. (1976) *The New World Primates*, Princeton University Press, Princeton.

Napier, J. R. and Napier, P. H. (1976) *A Handbook of Living Primates*, Academic Press, New York and London.

Napier, J. R. and Napier, P. H. (1970) *Old World Monkeys*, Academic Press, New York.

Rainer III, H. S. H. and Bourne, G. H. (1977) *Primate Conservation*, Academic Press, New York.

Schaller, G. B. (1963) *The Mountain Gorilla: Ecology and Behaviour*, University of Chicago Press, Chicago and London.

Short, R. V. and Weir, B. J. (eds) (1980) *The Great Apes of Africa*, Journals of Reproduction and Fertility, Colchester.

Simons, E. L. (1972) *Primate Evolution*, Collier Macmillan, London.

Struhsaker, T. T. (1975) *The Red Colobus Monkey*, University of Chicago Press, Chicago.

Sussman, R. W. (ed) (1979) *Primate Ecology: Problem-oriented Field Studies*, John Wiley, New York.

Szalay, F. S. and Delson, E. (1979) *Evolutionary History of the Primates*, Academic Press, New York.

GLOSSARY

Adaptation features of an animal which adjust it to its environment. Adaptations may be genetic, produced by evolution and hence not alterable within the animal's lifetime, or they may be phenotypic, produced by adjustment on the part of the individual and may be reversible within its lifetime. NATURAL SELECTION favors the survival of individuals whose adaptations adjust them better to their surroundings than other individuals with less successful adaptations.

Adaptive radiation the pattern in which different species develop from a common ancestor (as distinct from CONVERGENT EVOLUTION, the process whereby species from different origins became similar in response to the same SELECTIVE PRESSURES).

Adult a fully developed and mature individual, capable of breeding, but not necessarily doing so until social and/or ecological conditions allow.

Aerobic deriving energy from processes that require free atmospheric oxygen, as distinct from ANAEROBIC processes.

Agouti a grizzled coloration resulting from alternate light and dark barring of each hair.

Air sac a side-pouch of the larynx (the upper part of the windpipe), used in some primates and male walruses as a resonating chamber in producing calls.

Alloparent an animal behaving parentally toward infants that are not its own offspring; the shorthand jargon "HELPER" is most commonly applied to alloparents without any offspring of their own and it can be misleading if it is used to describe any non-breeding adults associated with infants, but which may or may not be "helping" by promoting their survival.

Alveolus a microscopic sac within the lungs providing the surface for gaseous exchange during respiration.

Anaerobic deriving energy from processes that do not require free oxygen, as distinct from AEROBIC processes.

Anal gland or sac a gland opening by a short duct either just inside the anus or on either side of it.

Anthropoid literally "man-like;" a member of the primate suborder Anthropoidea (monkeys, apes and man). Also, a great ape.

Aquatic living chiefly in water.

Arboreal living in trees.

Arteriole a small artery (ie muscular blood vessel carrying blood from the heart), eventually subdividing into minute capillaries.

Arterio-venous anastomosis (AVA) a connection between the ARTERIOLES carrying blood from the heart and the VENULES carrying it back to the heart.

Association a mixed-species group (polyspecific association) involving two or more species; relatively common among both Old and New World monkeys, but the most stable associations are found in forest-living guenons.

Axilla the angle between a forelimb and the body (in humans, the armpit).

Baculum (*os penis* or penis bone) an elongate bone present in the penis of certain mammals.

Binocular form of vision typical of mammals in which the same object is viewed simultaneously by both eyes; the coordination of the two images in the brain permits precise perception of distance.

Biomass a measure of the abundance of a life-form in terms of its mass, either absolute or per unit area (the population densities of two species may be identical in terms of the number of individuals of each, but due to their different sizes their biomasses may be quite different).

Biotic community a naturally occurring group of plants and animals in the same environment.

Bipedal walking on two legs. Only human beings exhibit habitual striding bipedalism. Some primate species may travel bipedally for short distances, and some (eg indri, bush babies, tarsiers) hop bipedally on the ground.

Blastocyst see IMPLANTATION.

Boreal region a zone geographically situated south of the Arctic and north of latitude 50°N: dominated by coniferous forest.

Brachiate to move around in the trees by arm-swinging beneath branches. In a broad sense all apes are brachiators, but only gibbons and siamangs exhibit a free-flight phase between hand-holds.

Brindled having inconspicuous dark streaks or flecks on a gray or tawny background.

Bursa (plural: bursae) a sac-like cavity (eg in ear of civets and Madagascan mongooses).

Cache a hidden store of food: also (verb) to hide for future use.

Canopy a fairly continuous layer in forests produced by the intermingling of branches of trees; may be fully continuous (closed) or broken by gaps (open). The crowns of some trees project above the canopy layer and are known as emergents.

Carnassial (teeth) opposing pair of teeth especially adapted to shear with a cutting (scissor-like) edge: in extant mammals the arrangement is unique to Carnivora and the teeth involved are the fourth upper premolar and first lower molar.

Catarrhine a "drooping-nosed" monkey, with nostrils relatively close together; term used for all Old World monkeys, apes and man in contrast to PLATYRRHINE monkeys of the New World.

Caudal gland an enlarged skin gland associated with the root of the tail. Subcaudal: placed below the root; supracaudal: above the root.

Cecum a blind sac in the digestive tract of a mammal, at the junction between the small and large intestines, particularly well developed in some specialized leaf-eaters.

Cerebral cortex the surface layer of cells (gray matter) covering the main part of the brain, consisting of the cerebral hemispheres.

Cerrado (central Brazil) a dry savanna region punctuated by patches of sparsely wooded vegetation.

Chaco (Bolivia and Paraguay) a lowland plains area containing soils carried down from the Andes; characterized by dry deciduous forest and scrub, transitional between rain forest and pampas grasslands.

Cheek pouch a pouch used for the temporary storage of food, found only in the typical monkeys of the Old World.

Chromatin material in the chromosomes of living cells containing the genes and proteins.

Class taxonomic category subordinate to a phylum and superior to an order (see TAXONOMY).

Clavicle the collar-bone.

Coniferous forest forest comprising largely evergreen conifers (firs, pines, spruces etc.), typically in climates either too dry or too cold to support deciduous forest. Most frequent in northern latitudes or in mountain ranges.

Consort, consortship in certain primates (eg Rhesus monkey, Savanna baboon, chimpanzees, orang-utan) males form temporary associations (consortships) with the females, ensuring priority of mating at the appropriate time.

Convergent evolution the independent acquisition of similar characters in evolution, as opposed to possession of similarities by virtue of descent from a common ancestor.

Crepuscular active in twilight.

Cryptic (coloration or locomotion) protecting through concealment.

Cusp a prominence on a cheek-tooth (premolar or molar).

Deciduous forest temperate and tropical forest with moderate rainfall and marked seasons. Typically, trees shed leaves during either cold or dry periods.

Delayed implantation see IMPLANTATION.

Den a shelter, natural or constructed, used for sleeping, for giving birth and raising young, and/or in winter; also the act of retiring to a den to give birth and raise young, or for winter shelter.

Dental formula a convention for summarizing the dental arrangement whereby the numbers of each type of tooth in each half of the upper and lower jaw are given; the numbers are always presented in the order: incisor (I), canine (C), premolar (P), molar (M). The final figure is the total number of teeth to be found in the skull. A typical example for Carnivora would be I3/3, C1/1, P4/4, M3/3 = 44.

Dentition the arrangement of teeth characteristic of a particular species.

Desert areas of low rainfall, typically with sparce scrub or grassland vegetation or lacking vegetation altogether.

Digit a finger or a toe.

Dichromatic in dichromatic species, males and females exhibit quite different color patterns (eg certain day-active lemurs, some New World monkeys, some Old World monkeys and certain gibbons).

Digital glands glands that occur between or on the toes.

Digitigrade method of walking on the toes without the heel touching the ground (cf PLANTIGRADE).

Dimorphism the existence of two distinct forms (polymorphism = several distinct forms); the term "sexual dimorphism" is applies to cases where the male and female of a species differ consistently in, for example, shape, size, coloration and armament.

Disjunct or **discontinuous distribution** geographical distribution of a species that is marked by gaps. Commonly brought about by fragmentation of suitable habitat, especially as a result of human intervention.

Dispersal the movements of animals, often as they reach maturity, away from their previous home range (equivalent to emigration). Distinct from **dispersion**, that is, the pattern in which things (perhaps animals, food supplies, nest sites) are distributed or scattered.

Display any relatively conspicuous pattern of behavior that conveys specific information to others, usually to members of the same species; can involve visual and/or vocal elements, as in threat, courtship or "greeting" displays.

Distal far from the point of attachment or origin (eg tip of tail).

Diurnal active in daytime.

Dominant see HIERARCHY.

Dormancy a period of inactivity; many bears, for example, are dormant for a period in winter: this is not true HIBERNATION, as pulse rate and body temperature do not drop markedly.

Ecology the study of plants and animals in relation to their natural environmental setting. Each species may be said to occupy a distinctive ecological NICHE.

Ecosystem a unit of the environment within which living and nonliving elements interact.

Ecotype a genetic variety within a single species, adapted for local ecological conditions.

Elongate relatively long (eg of canine teeth, longer than those of an ancestor, a related animal, or than adjacent teeth).

Emigration departure of animal(s), usually at or about the time of reaching adulthood, from the group or place of birth.

Enzootic concerning disease regularly found within an animal population (endemic applies specifically to people) as distinct from EPIZOOTIC.

Epizootic a disease outbreak in an animal population at a specific time (but not persistently, as in ENZOOTIC); if an epizootic wave of infection eventually stabilizes in an area, it becomes enzootic.

Erectile capable of being raised to an erect position (erectile mane).

Esophagus the gullet connecting the mouth with the stomach.

Estrus the period in the estrous cycle of female mammals at which they are often attractive to males and receptive to mating. The period coincides with the maturation of eggs and ovulation (the release of mature eggs from the ovaries). Animals in estrus are often said to be "on heat" or "in heat." In primates, if the egg is not fertilized the subsequent degeneration of uterine walls (endometrium) leads to menstrual bleeding. In some species ovulation is triggered by copulation and this is called **induced ovulation**, as distinct from spontaneous ovulation.

Exudate natural plant exudates include gums and resins; damage to plants (eg by marmosets) can lead to loss of sap as well. Certain PROSIMIANS and other primates (eg marmosets) rely heavily on exudates as a source of food.

Family a taxonomic division subordinate to an order and superior to a genus (see TAXONOMY).

Feces excrement from the bowels: colloquially known as droppings or scats.

Feral living in the wild (of domesticated animals, eg cat, dog).

Fermentation the decomposition of organic substances by microorganisms. In some mammals, parts of the digestive tract (eg the cecum) may be inhabited by bacteria that break down cellulose and release nutrients.

Fissipedia (suborder) name given by some taxonomists to modern terrestrial carnivores to distinguish them from the suborder Pinnipedia which describes the marine carnivores. Here we treat both as full orders: the Carnivora and the Pinnipedia.

Fitness a measure of the ability of an animal (with one genotype or genetic make-up) to leave viable offspring in comparison to other individuals (with different genotypes). The process of NATURAL SELECTION, often called survival of the fittest, determines which characteristics have the greatest fitness, ie are most likely to enable their bearers to survive and rear young which will in turn bear those characteristics. (See INCLUSIVE FITNESS.)

Flehmen German word describing a facial expression in which the lips are pulled back, head often lifted, teeth sometimes clapped rapidly together and nose wrinkled. Often associated with animals (especially males) sniffing scent marks or socially important odors (eg scent of estrous female). Possibly involved in transmission of odor to JACOBSON'S ORGAN.

Folivore an animal eating mainly leaves.

Follicle a small sac, therefore (a) a mass of ovarian cells that produces an ovum, (b) an indentation in the skin from which hair grows.

Forestomach a specialized part of the stomach consisting of two compartments (presaccus and saccus).

Fossorial burrowing (of life-style or behavior).

Frugivore an animal eating mainly fruits.

Furbearer term applied to mammals whose pelts have commercial value and form part of the fur harvest.

Gallery forest luxuriant forest lining the banks of watercourses.

Gamete a male or female reproductive cell (ovum or spermatozoon).

Gene the basic unit of heredity: a portion of DNA molecule coding for a given trait and passed, through replication at reproduction, from generation to generation. Genes are expressed as ADAPTATIONS and consequently are the most fundamental units (more so than individuals) on which NATURAL SELECTION acts.

Generalist an animal whose life-style does not involve highly specialized strategems (cf SPECIALIST): for example, feeding on a variety of foods which may require different foraging techniques.

Genus (plural genera) a taxonomic division superior to species and subordinate to family (see TAXONOMY).

Gestation the period of development within the uterus; the process of **delayed implantation** can result in the period of pregnancy being longer than the period during which the embryo is actually developing (See also IMPLANTATION.)

Glands (marking) specialized glandular areas of the skin, used in depositing SCENT MARKS.

Great call a protracted series of notes, rising to a climax, produced by the female as part of the group song in lesser apes.

Grizzled sprinkled or streaked with gray.

Harem group a social group consisting of a single adult male, at least two adult females and immature animals; the most common pattern of social organization among mammals.

Helper jargon for an individual, generally without young of its own, which contributes to the survival of the offspring of others by behaving parentally toward them (see ALLOPARENT).

Hemoglobin an iron-containing protein in the red corpuscles which plays a crucial role in oxygen exchange between blood and tissues in mammals.

Herbivore an animal eating mainly plants or parts of plants.

Hierarchy (social or dominance) the existence of divisions within society, based on the outcome of interactions which show some individuals to be consistently dominant to others. Higher-ranking individuals thus have control of aspects (eg access to food or mates) of the life and behavior of low-ranking ones. Hierarchies may be branching, but simple linear ones are often called peck orders (after the behavior of farmyard chickens).

Higher primate one of the more advanced primates, known as ANTHROPOIDS.

Holarctic realm a region of the world including North America, Greenland, Europe, and Asia apart from the southwest, southeast and India.

Home range the area in which an animal normally lives (generally excluding rare excursions or migrations), irrespective of whether or not the area is defended from other animals (cf TERRITORY).

Hybrid the offspring of parents of different species.

Hyoid bones skeletal elements in the throat region, supporting the trachea, larynx and base of the tongue (derived in evolutionary history from the gill arches of ancestral fishes).

Implantation the process whereby the free-floating blastocyst (early embryo) becomes attached to the uterine wall in mammals. At the point of implantation a complex network of blood vessels develops to link mother and embryo (the PLACENTA). In **delayed implantation** the blastocyst remains dormant in the uterus for periods varying, between species, from 12 days to 11 months. Delayed implantation may be obligatory or facultative and is known for some members of the Carnivora and Pinnipedia.

Inclusive fitness a measure of the animal's FITNESS which is based on the number of genes, rather than the number of its offspring, present in subsequent generations. This is a more complete measure of fitness, since it incorporates the effect of, for example, alloparenthood, wherein individuals may help to rear the offspring of their relatives (see KIN SELECTION: ALLOPARENT).

Induced ovulation see ESTRUS.

Infanticide the killing of infants. Infanticide has been recorded notably in species in which a bachelor male may take over a HAREM from its resident male(s).

Insectivore an animal eating mainly arthropods (insects, spiders).

Ischial callosities specialized, hardened pads of tissue present on the buttocks of some monkeys and apes. Each overlies a flattened projection of the ischium bone of the pelvis. Known also as "sitting pads," they are found in Old World monkeys and lesser apes.

Jacobson's organ a structure in a foramen (small opening) in the palate of many vertebrates which appears to be involved is olfactory communication. Molecules of scent may be sampled in these organs.

Juvenile no longer possessing the characteristics of an infant, but not yet fully adult.

Kin selection a facet of NATURAL SELECTION whereby an animal's fitness is affected by the survival of its relatives or kin. Kin selection may be the process whereby some ALLOPARENTAL behavior evolved: an individual behaving in a way which promotes the survival of its kin increases its own INCLUSIVE FITNESS, despite the *apparent* selflessness of its behavior.

Knuckle-walk to walk on all fours with the weight of the front part of the body carried on the knuckles; found only in gorillas and chimpanzees.

Lactation (verb: lactate) the secretion of milk from MAMMARY GLANDS.

Lanugo the birth-coat of mammals which is shed to be replaced by the adult coat.

Latrine a place where FECES are regularly left (often together with other SCENT MARKS); associated with olfactory communication.

Lesser apes the gibbons and siamang.

Liana a climbing plant. In rain forests large numbers of often woody, twisted lianas hang down like ropes from the crowns of trees.

Lower primate one of the more primitive primates known as PROSIMIANS.

Lumbar a term locating anatomical features in the loin region, eg lumbar vertebrae are at the base of the spine.

Mammal a member of a CLASS of VERTEBRATE animals having MAMMARY GLANDS which produce milk with which they nurse their young (properly: Mammalia).

Mammary glands glands of female mammals that secrete milk.

Mangrove forest tropical forest developed on sheltered muddy shores of deltas and estuaries exposed to tide. Vegetation is almost entirely woody.

Mask colloquial term for the face of a mammal, especially a dog, fox or cat.

Matriline a related group of animals linked by descent through females alone.

Melanism darkness of color due to presence of the black pigment melanin.

Menstrual cycle an approximately monthly cycle involving alternation of ovulation and menstruation (loss or blood from the vulva at monthly intervals) until pregnancy intervenes; found in humans, great apes, Old World monkeys and, to varying degrees, in New World monkeys.

Metabolic rate the rate at which the chemical processes of the body occur.

Migration movement, usually seasonal, from one region or climate to another for purposes of feeding or breeding.

Monogamy a mating system in which individuals have only one mate per breeding season.

Mutation a structural change in a gene which can thus give rise to a new heritable characteristic.

Myoglobin a protein related to HEMOGLOBIN, found in the muscles of vertebrates; like hemoglobin, it is involved in the oxygen exchange processes of respiration.

Myopia short-sightedness.

Nasolacrimal duct a duct or canal between the nostrils and the eye.

Natal range the home range into which an individual was born (natal = of or from one's birth).

Natural selection the process whereby individuals with the most appropriate ADAPTATIONS are more successful than other individuals, and hence survive to produce more offspring. To the extent that the successful traits are heritable (genetic) they will therefore spread in the population.

Niche the role of a species within the community, defined in terms of all aspects of its life-style (eg food, competitors, predators, and other resource requirements).

Nocturnal active at nighttime.

Olfaction, olfactory the olfactory sense is the sense of smell, depending on receptors located in the epithelium (surface membrane) lining the nasal cavity.

Omnivore an animal eating a varied diet including both animal and plant tissue.

Opportunist (of feeding) flexible behavior of exploiting circumstances to take a wide range of food items: characteristic of many species of Carnivora. See GENERALIST; SPECIALIST.

Order a taxonomic division subordinate to class and superior to family (see TAXONOMY).

Ovulation (verb: ovulate) the shedding of mature ova (eggs) from the ovaries where they are produced (see ESTRUS).

Papilla (plural: papillae) a small nipple-like projection.

Parturition the process of giving birth (hence *post partum*—after birth).

Patagium a gliding membrane typically stretching down the sides of the body between the fore-and hindlimbs and perhaps including part of the tail. Found in colugos, flying squirrels, bats etc.

Pelvis a girdle of bones that supports the hindlimbs of vertebrates.

Perineal glands glandular tissue occurring between the anus and genitalia.

Perineal swelling a swelling of the naked area of skin in the region of the anus and vulva of a female primate, found in chimpanzees and some Old World monkeys.

Pinna (plural: pinnae) the projecting cartilaginous portion of the external ear.

Placenta, placental mammals a structure that connects the fetus and the mother's womb to ensure a supply of nutrients to the fetus and removal of its waste products. Only placental mammals have a well-developed placenta; marsupials have a rudimentary placenta or none, and monotremes lay eggs.

Plantigrade way of walking on the soles of the feet, including the heels (cf DIGITIGRADE).

Platyrrhine a "flat-nosed" monkey with widely separated nostrils. Term commonly used for all New World monkeys in contrast to CATARRHINE monkeys of the Old World.

Polyandrous see POLYGYNOUS.

Polygamous a mating system wherein an individual has more than one mate per breeding season.

Polygynous a mating system in which a male mates with several females during one breeding season (as opposed to polyandrous, where one female mates with several males).

Population a more or less separate (discrete) group of animals of the same species within a given BIOTIC COMMUNITY.

Predator an animal which forages for live prey; hence "anti-predator behavior" describes the evasive actions of the prey.

Prehensile capable of grasping (eg of the tail).

Primary forest forest that has remained undisturbed for a long time and has reached a mature (climax) condition; primary rain forest may take centuries to become established.

Primate a member of an order comprising the apes, monkeys and related forms, including man.

Process (anatomical) an outgrowth or protuberance.

Promiscuous a mating system wherein an individual mates more or less indiscriminately.

Prosimian literally "before the monkeys"; a member of the relatively primitive primate suborder Prosimii (lemurs, lorises and tarsiers).

Proximal near to the point of attachment or origin (eg the base of the tail).

Puberty the attainment of sexual maturity. In addition to maturation of the primary sex organs (ovaries, testes), primates may exhibit secondary sexual characteristics at puberty. Among higher primates it is usual to find a growth spurt at the time of puberty in males and females.

Quadrumanous using both hands and feet for grasping.

Quadrupedal walking on all fours, as opposed to walking on two legs (BIPEDAL) or moving suspended beneath branches in trees (SUSPENSORY MOVEMENT).

Race a taxonomic division subordinate to SUBSPECIES but linking populations with similar distinct characteristics.

Radiation see ADAPTIVE RADIATION

Radio-tracking a technique used for monitoring an individual's movements remotely; it involves affixing a radio transmitter to the animal and thereafter receiving a signal through directional antennae, which enables the subject's position to be plotted. The transmitter is often attached to a collar, hence "radio-collar."

Rain forest tropical and subtropical forest with abundant and year-round rainfall. Typically species rich and diverse.

Receptive state of a female mammal ready to mate or in ESTRUS.

Reduced (anatomical) of relatively small dimension (eg of certain bones, by comparison with those of an ancestor or related animals).

Reproductive rate the rate of production of offspring; the net productive rate may be defined as the average number of female offspring produced by each female during her entire lifetime.

Retractile (of claws) able to be withdrawn into protective sheaths.

Rhinarium a naked area of moist skin surrounding the nostrils in many mammals.

Savanna tropical grasslands of Africa, Central and South America and Australia. Typically on flat plains and plateaux with seasonal pattern of rainfall. Three categories—*savanna woodland, savanna parkland* and *savanna grassland*—represent a gradual transition from closed woodland to open grassland.

Scapula the shoulder-blade. Primates typically have a mobile scapula in association with their versatile movements in the trees.

Scent gland an organ secreting odorous material with communicative properties; see SCENT MARK.

Scent mark a site where the secretions of scent glands, or urine or FECES, are deposited and which has communicative significance. Often left regularly at traditional sites which are also visually conspicuous. Also the "chemical message" left by this means; and (verb) to leave such a deposit.

Scrub a vegetation dominated by shrubs—woody plants usually with more than one stem. Naturally occurs most often on the arid side of forest or grassland types, but is often artificially created by man as a result of forest destruction.

Seasonality (of births) the restriction of births to a particular time of the year.

Sebaceous gland secretory tissue producing oily substances, for example lubricating and waterproofing hair, or specialized to produce odorous secretions.

Secondary forest (or growth) regenerating forest that has not yet reached the climax condition of PRIMARY FOREST.

Sectorial premolar one of the front lower premolars of Old World monkeys and apes, specially adapted for shearing against the rear edge of the upper canine.

Selective pressure a factor affecting the reproductive success of individuals (whose success will depend on their FITNESS, ie the extent to which they are adapted to thrive under that selective pressure).

Septum a partition separating two parts of an organism. The nasal septum consists of a fleshy part separating the nostrils and a vertical, bony plate dividing the nasal cavity.

Simian (literally "ape-like") a monkey or ape. Often used as a synonym of ANTHROPOID or HIGHER PRIMATE.

Sinus a cavity in bone or tissue.

Solitary living on its own, as opposed to social or group-living in life-style.

Specialist an animal whose life-style involves highly specialized stratagems: (eg feeding with one technique on a particular food).

Species a taxonomic division subordinate to genus and superior to SUBSPECIES. In general a species is a group of animals similar in structure and which are able to breed and produce viable offspring. See TAXONOMY.

Speciation the process by which new SPECIES arise in evolution. It is widely accepted that it occurs when a single-species population is divided by some geographical barrier.

Steppe open grassy plains of the central temperate zone of Eurasia or North America (prairies), characterized by low and sporadic rainfall and a wide annual temperature variation. In cold steppe, temperatures drop well below freezing point in winter, with rainfall concentrated in the summer or evenly distributed throughout year, while in hot steppe, winter temperatures are higher and rainfall concentrated in winter months.

Subadult no longer an infant or juvenile but not yet fully adult physically and/or socially.

Subfamily a division of a FAMILY.

Subfossil an incompletely fossilized specimen from a recent species.

Sublingua or **subtongue** a flap of tissue beneath the tongue in mammals, retained in most primates though vestigial in New World monkeys; particularly in lemurs and lorises.

Suborder a subdivision of an ORDER.

Subordinate see HIERARCHY.

Subspecies a recognizable subpopulation of a single SPECIES, typically with a distinct geographical distribution.

Surplus killing a phenomenon where more (sometimes very many more) prey are killed than can immediately be consumed by the killer or its companions.

Suspensory movement movement through the trees by hanging and swinging beneath, rather than running along the tops of branches. See also BRACHIATE.

Taiga northernmost coniferous forest, with open boggy/rocky areas in between.

Tapetum lucidum a reflecting layer located behind the retina of the eye, commonly found in nocturnal mammals.

Taxonomy the science of classifying organisms. It is very convenient to group together animals which share common features and are thought to have common descent. Each individual is thus a member of a series of ever-broader categories (individual—species—genus—family—order—class—phylum) and each of these can be further divided where it is convenient (eg subspecies, superfamily or infraorder). The SPECIES is a convenient unit in that it links animals according to an obvious criterion, namely that they interbreed successfully. However, the unit on which NATURAL SELECTION operates is the individual: it is by the differential reproductive success of individuals bearing different characteristics that evolutionary change proceeds.

Terrestrial living on land.

Territory an area defended from intruders by an individual or group. Originally the term was used where ranges were exclusive and obviously defended at their borders. A more general definition of territoriality allows some overlap between neighbors by defining territoriality as a system of spacing wherein home ranges do not overlap randomly—that is, the location of one individual's or group's home range influences that of others.

Testosterone a male hormone synthesized in the testes and responsible for the expression of many male characteristics (contrast the female hormone estrogen produced in the ovaries).

Tooth-comb a dental modification in which the incisor teeth form a comb-like structure.

Tubercle a small rounded projection or nodule (eg of bone).

Tundra barren treeless lands of the far north of Eurasia and North America, on mountain tops and Arctic islands. Vegetation is dominated by low shrubs, herbaceous perennials, with mosses and lichens.

Underfur the thick soft undercoat fur lying beneath the longer and coarser guard hairs.

Vector an individual or species which transmits a disease.

Ventral on the lower or bottom side or surface: thus ventral abdominal glands occur on the underside of the abdomen.

Venule a small tributary conveying blood from the capillary bed to a vein (cf ARTERIOLE.)

Vertebrate an animal with a backbone: a division of the phylum Chordata which includes animals with notochords (as distinct from invertebrates).

Vestigial a characteristic with little or no contemporary use, but derived from one which was useful and well-developed in an ancestral form.

Vibrissae stiff, coarse hairs richly supplied with nerves, and with a sensory (tactile) function, found especially around the snout.

Xerophytic forest a forest found in areas with relatively low rainfall. Xerophytic plants are adapted to protect themselves against browsing (eg well-developed spines) and to limit water loss (eg small, leathery leaves, often with a waxy coating).

INDEX

A **bold number** indicates a major section of the main text, following a heading; a ***bold italic*** number indicates a fact box on a single species; a single number in (parentheses) indicates that the animal name or subjects are to be found in a boxed feature and a double number in (parentheses) indicates that the animal name or subjects are to be found in a spread special feature. *Italic* numbers refer to illustrations.

A

Allenopithecus
nigroviridis ***88*** (*see* monkey, Allen's swamp)
Allocebus
trichotis ***35*** (*see* lemur, Hairy-eared dwarf)
Alouatta
belzebul ***65*** (*see* howler, Red-handed)
caraya ***65*** (*see* howler, Black)
fusca ***65***
palliata ***65*** (*see* howler, Mantled)
seniculus ***65***
villosa ***65***
Ameslan 21
Anathana
ellioti ***148***
anatomy
of colobine monkeys 102
of primates *12-13*
of tree shrews 144-146
of typical monkeys **74-76**
angwantibo *14*, 38, *39*, *40*, ***41***
anthropoids 10, 13
anti-predator behavior
in angwantibos 38
in pottos 37-38
see also defense
Aotus
trivirgatus ***62*** (*see* monkey, Night)
ape
Barbary ***89*** (*see* macaque, Barbary)
Rock ***89*** (*see* macaque, Barbary)
see also apes
apes **116-117**
great 13, 116
lesser 13, 116
see also ape
Archaeoindris (26)
Archaeolemur (26)
Arctocebus
calabarensis ***41*** (*see* angwantibo)
associations
Saddle-back–Emperor tamarin (52-53)
Spider monkey–capuchin, Squirrel monkey–uakari 60
Squirrel monkey–capuchin 60-61
Ateles
belzebuth ***65***
fusciceps ***65***
geoffroyi ***65*** (*see* monkey, Black-handed spider)
paniscus ***65***
aunting 20
see also baby-sitting, helpers, infant care
Australopithecus 13
avahi ***35*** (*see* lemur, Woolly)
Avahi
laniger ***35*** (*see* lemur, Woolly)
aye-aye 22-23, *23*, (31), *31*, ***35***

B

babakoto 31
baboon *13*, 45, 74, 75, 76, *76-77*, 78, **79**, *80*, 81, 82, 83, 84
Chacma *77*, ***89*** (*see also* baboon, Savanna)
Common ***89*** (*see* baboon, Savanna)
Gelada ***89*** (*see* gelada)
Guinea *77*, *71*, ***81***
Hamadryas *18*, 45, *77*, 79, ***89***, (98-99), *98-99*
Olive *75*, *77*, ***89***, (96-97), *96-97* (*see also* baboon, Savanna)
Savanna 44, *77*, *76*, 78, 79, 80, ***89*** (*see also* baboon, Olive; baboon, Yellow)
Yellow *19*, *77*, ***89*** (*see also* baboon, Savanna)
baby-sitting 105-106
see also aunting, helpers, infant care
bands
all-male 107, 114
Hamadryas baboon *98*, 98-99
barking
in Gray-cheeked mangabeys 95
beeloh ***120*** (*see* gibbon, Kloss)
bonobo ***126*** (*see* chimpanzee, Pygmy)
brachiators 15
true 116, *118-119*
Brachyteles
arachnoides ***65*** (*see* monkey, Woolly spider)
brains
primate *12*, 13, 14
buffering, agonistic
in Barbary macaques 92-93
bush baby *19*, 23, *37*, 38-41
Allen's *14*, ***40***, 41
Demidoff's ***40*** (*see* bush baby, Dwarf)
Dwarf *14*, *38*, 38, ***40***, 40
Lesser *36*, ***40***, *41*
Needle-clawed *14*, *23*, 38, *39*, ***41***
Senegal ***40*** (*see* bush baby, Lesser)
Thick-tailed *22*, *37*, 38, *39*, ***40***
buttocks
monkey 44

C

Cacajao
melanocephalus ***65***
rubicundus ***65*** (*see* uakari, Red)
Callicebus
moloch ***62*** (*see* titi, Dusky)
personatus ***64*** (*see* titi, Masked)
torquatus ***62*** (*see* titi, Yellow-handed)
Callimico
goeldii ***51*** (*see* monkey, Goeldi's)
Callithrix
argentata ***50*** (*see* marmoset, Bare-ear; marmoset, Black-tailed; marmoset, Silvery)
aurita ***50*** (*see* marmoset, Buffy tufted-ear)
flaviceps ***50*** (*see* marmoset, Buffy-headed)
geoffroyi ***50***
humeralifer ***50*** (*see* marmoset, Tassel-ear)
jacchus ***50***
penicillata ***50*** (*see* marmoset, Black tufted ear)
callitrichids (Callitrichidae) 44-45, **46-49**
callosites, ischial
in colobine monkeys 102
in Old World monkeys 44
calls
in forest monkeys 94
in mangabeys (94-95), *94-95*
in orang-utans (134)
see also howling, vocalizations
capuchin 45, 59, 60-61, 61
Black-capped ***64*** (*see* capuchin, Brown)
Brown 56, 57, 58, 59, *59*, (60), *60*, 60, *61*, ***64***
Tufted ***64*** (*see* capuchin, Brown)
Wedge-capped (Weeper capuchin) ***64***
Weeper ***64***
White-faced ***64***
White-fronted 56, 57, 59, 60, ***64***
White-throated (White-faced capuchin) ***64***
catarrhines 44, *45*
cebids (Cebidae) 45, **56-61**
Cebuella
pygmaea ***50*** (*see* marmoset, Pygmy)
Cebus
albifrons ***64*** (*see* capuchin, White-fronted)
apella ***64*** (*see* capuchin, Brown)
capucinus ***64***
nigrivittatus ***64***
Cercocebus 80
albigena ***86*** (*see* mangabey, Gray-cheeked)
aterrimus ***86***
galeritus ***86*** (*see* mangabey, Agile)
torquatus ***86*** (*see* mangabey, White)
Cercopithecidae 45
cercopithecines (Cercopithecinae) 45, **74-85**
Cercopithecus 80
aethiops ***86*** (*see* vervet)
ascanius ***86*** (*see* monkey, Redtail)
campbelli ***88***
cephus ***86*** (*see* monkey, Moustached)
diana ***88*** (*see* monkey, Diana)
dryas ***88***
erythrogaster ***87***
erythrotis ***87***
hamlyni ***87***
l'hoesti ***87***
mitis ***88*** (*see* monkey, Blue)
mona ***88***
neglectus ***88*** (*see* monkey De Brazza's)
nictitans ***88***
petaurista ***87***
pogonias ***88***
preussi—*see C. l'hoesti*
wolfi ***88***
cheek pouches
in typical monkeys 75
Cheirogaleidae 22, **30-31**
Cheirogaleus
major ***35*** (*see* lemur, Greater dwarf)
medius ***35*** (*see* lemur, Fat-tailed dwarf)
chimpanzee *10*, *14*, *19*, 82, 116, 117, *126-131*, 126-131, (128), (130)
Common *116*, ***116***, 126-127, 128, 129
Pygmy ***126***, 126, *127*, 127, 128, 129
Chiropotes
albinasus ***64***
satanus ***64***
clans
Hamadryas baboon *98-99*, 98-99
claws
needle 31
toilet 42, *43*
colobin
Eastern black-and-white ***112*** (*see* guereza)
Colobinae 45, 72, **102-109**
colobus 102, 104, 106, 108
Angolan black-and-white (White-epauletted black colobus) ***111***
Bay ***112*** (*see* colobus, Guinea forest red)
Cameroon red ***112***
Guinea forest black 16, 103, 108, ***112***
Guinea forest red 16, *102*, 106, ***112***
Magistrate black ***112*** (*see* guereza)
Olive 103, 104, ***112***
Pennant's red 103, 106, 107, ***112***
Preuss's red (Cameroon red colobus) ***112***
Regal black ***112*** (*see* colobus, Guinea forest black)
Satanic black 104, 106, ***111***
Tana River red ***112***
Ursine black ***112*** (*see* colobus, Guinea forest black)
Van Beneden's ***112*** (*see* colobus, Olive)
Western black-and-white ***112*** (*see* colobus, Guinea forest black)
White-epauletted black ***111***
White-mantled black ***112*** (*see* guereza)
Colobus 45, 104
angolensis ***111***
guereza ***112*** (*see* guereza)
polykomos ***112*** (*see* colobus, Guinea forest black)
satanas ***111*** (*see* colobus, Satanic black)
colugo **150-151**, *151-152*
Malayan ***150***, *150-151*, 150
Philippine ***150***, 150
comb tines
in flying lemurs 151
competition
cebid monkey 57-58
conservation
of gorillas 143
of typical monkeys **85**
see also energy conservation
Cynocephalidae 150
Cynocephalus
variegatus ***150*** (*see* colugo, Malayan)
volans ***150*** (*see* colugo, Philippine)

D

Daubentonia
madagascariensis ***35*** (*see* aye-aye)
robusta (31)
defense, mutual
in Brown capuchins (60)
see also anti-predator behavior
Dendrogale
melanura ***148***
murina ***148*** (*see* tree shrew, Northern smooth-tailed)
dentition
procumbent 31
dermopterans (Dermoptera) **150-151**
digestion
in colobine monkeys 72, 103-104
in flying lemurs 151
in hindgut fermenters 72
disease
in baboons 85
in orang-utans 135
in typical monkeys 85
Dolichopithecus 102
dominance
in Barbary macaques 92-93
in Brown capuchins (60)
in chimpanzees (130), *130*
in Mantled howlers 73
in matrilinear baboons *20*, 20
in matrilinear macaques *20*, 20
in Olive baboons 96-97, *97*
in Rhesus macaques 91
dormancy
in dwarf lemurs 31
douc 104, 108
Black-footed ***110*** (*see* douc, Black-shanked)
Black-shanked 107, ***110***
Red-shanked 104, ***110***
douroucouli ***62*** (*see* monkey, Night)
drill *76*, 79, ***89***

E

endangered species
Buffy-headed marmosets *48*
cebid monkeys 61
Golden monkeys (109)
indri and sifakas 33
Lion tamarins (54-55)
Mountain gorillas *137*, 142
primates 18
typical monkeys 85
Woolly spider monkeys *61*
energy conservation
in Mantled howlers 73
Erythrocebus
patas ***88*** (*see* monkey, Patas)
Euoticus—*see* bush baby, Needle-clawed
extinction
of *Daubentonia robusta* (31)
of indri and sifakas 33
of lemurs 23, 24, (26), 31
eyeshine
in angwantibos *40*
see also tapetum lucidum

F

faces
chimpanzee *21*, *127*
macaque *78-79*
mandrill *44*
primate 13
families
gibbon 122, 123
Hamadryas baboon *98*, 98-99
feeding behavior
in Allen's bush babies *14*
in angwantibos *14*
in apes 116
in Brown capuchins (60)
in chimpanzees *14*
in colobine monkeys 103, 106
in Dwarf bush babies *14*
in gibbons 121
in gorillas *14*, *16-17*
in Himalayan Rhesus macaques 90
in mandrills *14*
in Mantled howlers 72, *73*, 73
in Needle-clawed bush babies *14*
in pottos *14*
in primates 15-16
in Red colobus monkeys *14*
in Saddle-back and Emperor tamarins 53
in Squirrel monkeys *15*
in typical monkeys **80-81**
see also gum-feeding
feet
baboon *13*
orang-utan *13*
primate 12, 15
siamang *13*
tamarin *13*
typical monkey 75

fighting
in Brown capuchins 59
in chimpanzees (130-131), *130-131*
in gorillas 141, 142
in Hamadryas baboons *98-99*, 99
in Night monkeys 68
in orang-utans 134
in Rhesus macaques 91
folivores 16
friendship
in Olive baboons 97
frugivores 16
furs
colobine monkey 108

G

Galaginae 37
galago—*see* bush baby
Galago
alleni ***40*** (*see* bush baby, Allen's)
crassicaudatus ***40*** (*see* bush baby, Thick-tailed)
demidovii ***40*** (*see* bush baby, Dwarf)
elegantulus ***41*** (*see* bush baby, Needle-clawed)
inustus ***41*** (*see* bush bay, Neddle-clawed)
senegalensis ***40*** (*see* bush baby, Lesser)
Galeopithecidae 150
gelada *18*, *19*, *76*, 76, 78, 79, 81, 85, ***89***
gibbon *10*, *13*, *19*, 116, 117, **118-123**
Agile ***120***, *120-121*, 121, 122, 123
Capped ***120*** (*see* gibbon, Pileated)
Common ***120*** (*see* gibbon, Lar)
Concolor ***120***, *120-121*, 121, 123
Crested ***120*** (*see* gibbon, Concolor)
Dwarf ***120*** (*see* gibbon, Kloss)
Gray ***120*** (*see* gibbon, Müller's)
Hoolock ***120***, *120*, 121, 123
Kloss ***120***, *120*, 121, 122, 123, (124-125), *124-125*
Lar *118-119*, ***120***, *120*, 121, 122, 123
Moloch ***120***, *120*, 121, *122*, 122, 123
Müller's ***120***, *121*, 121, 123
Pileated ***120***, *120*, 121, 122, 123
Silvery ***120*** (*see* gibbon, Moloch)
White-browed ***120*** (*see* gibbon, Hoolock)
White-cheeked ***120*** (*see* gibbon, Concolor)
White-handed ***120*** (*see* gibbon, Lar)
gliding
in flying lemurs 150, *151-152*
Gloger's Rule 148
gorilla *10*, *11*, *12*, *13*, *14*, *16-17*, *19*, *116*, 116, 117, ***136***, **136-143**, *136-143*
Gorilla
gorilla ***136*** (*see* gorilla)
grivet ***86*** (*see* vervet)
grooming
in Barbary macaques *92*, 92
in cebid monkeys 61
in gorillas *140*, 141
in Redtail monkeys *101*
in Rhesus macaques *21*
in titis *70*, 71
guenon *12*, *74*, 75, 76, *77*, 78, **80**, 82, 83, 84-85
Crowned ***88***
Schmidt's ***78*** (*see* monkey, Redtail)
guereza 103, 104, *106-107*, 107, 108, ***112***
gum-feeding
in dwarf lemurs 31
in marmosets (48)
in Pygmy marmosets *47*

H

habitat destruction
for cebid monkeys 61
for chimpanzees 130
for colobine monkeys 108
for gibbons 123
for Golden monkeys (109)
for gorillas 142
for Lion tamarins 54, 55
for orang-utans 134, 135
for primates 16-18
for typical monkeys 85
Hadropithecus (26)
hands
baboon 81
gibbon *13*
gorilla *13*
macaque *13*
primate 12-13, 15
Spider monkey *13*
typical monkey 75
Hapalemur
griseus ***35***
griseus alaotrensis ***35***
griseus griseus ***35*** (*see* lemur, Gray gentle)
griseus occidentalis ***35***
simus ***35***
Hapalemurinae (gentle lemurs) 24, 25, 26, 28
harems
Blue and Redtail monkey 100
Gelada baboon *18*, *19*
gorilla *19*, 117, 141, 142
Hamadryas baboon *18*, 98-99, *98-99*
howler and capuchin 59
primate 19, 20
helpers
Squirrel monkey 60
hocheur (Spot-nosed monkey) ***88***
Hominoidea 116
Homo
sapiens 13
hooting
in Night monkeys 68
howler 45, 59, 61, 115
Black *57*, *58-59*, ***65***
Brown ***65***
Guatemalan (Mexican black howler) ***65***
Mantled 57, ***65***, (72-73), *72-73*
Mexican black ***65***
Red ***65***
Red-handed *45*, ***65***
howling
in howle monkeys *72*, 72, 73
in indri 33
huddling
in Barbary macaques *92*, 92
hunting
for gorillas *142*, 142-143
see also hunting, cooperative
hunting, cooperative
in baboons 81
in chimpanzees 128
hybridization
Redtail–Blue monkey (100-101), *100*
Hylobates 121
agilis ***120*** (*see* gibbon, Agile)
concolor ***120*** (*see* gibbon, Concolor)
hoolock ***120*** (*see* gibbon, Hoolock)
klossi ***120*** (*see* gibbon, Kloss)
lar ***120*** (*see* gibbon, Lar)
moloch ***120*** (*see* gibbon, Moloch)
muelleri ***120*** (*see* gibbon, Müller's)
pileatus ***120*** (*see* gibbon, Pileated)
syndactylus ***120*** (*see* siamang)
Hylobatidae 116

I

indri *22*, 31, *32*, 32, 33, ***35***
Indri
indri ***35*** (*see* indri)
indriids (Indriidae) 22, **31-33**
indris ***35*** (*see* indri)
woolly ***35*** (*see* lemur, Woolly)
infant care
male 20, 49, 71-72, (92-93), *92*, 97, 121-122
maternal 20, 83
shared 60, 105-106
infanticide
in chimpanzees 131
in gorillas 142
in Hanuman langurs (114-115)
in monkeys 115
in primates 20
insect prey
of chimpanzees (128)
of tree shrews *145*, *149*
intelligence
in gorillas 136
in orang-utans 135
interbreeding—*see* hybridization

K

killing techniques
of tarsiers 42

L

Lagothrix
flavicauda ***65***
lagotricha ***65*** (*see* monkey, Humboldt's woolly)
language
and chimpanzees 126
and gorillas 136
and primates 21
langur
Common ***111*** (*see* langur Hanuman)
Hanuman 103, 104, *105-106*, 107, 108, *109*, 109, ***111***, *112-113*, (114-115), *114-115*
Malabar ***111***
Moupin ***110*** (*see* monkey, Golden)
Nilgiri ***111*** (*see* monkey, Hooded black leaf)
Pagai Island ***110*** (*see* monkey, Pig-tailed snub-nosed)
Pig-tailed snub-nosed ***110*** (*see* monkey, Pig-tailed snub-nosed)
lemur
Alaotran gentle ***35***
Black *24*, 24, ***34***
Black-and-white ruffed ***34***
Broad-nosed gentle ***35***
Brown *25*, 26, ***34***
Brown lesser mouse 30, *31*, ***35***
Collared 26, ***34***
Coquerel's mouse 30, 31, ***35***
Crowned 25, 26, ***34***
Fat-tailed dwarf *30*, 30-31, ***35***
Fork-crowned dwarf 31, ***35***
Gray gentle *25*, ***35***
Gray lesser mouse 30, ***35***
Gray-backed sportive ***35***
Greater dwarf 30-31, ***35***
Hairy-eared dwarf 30, 31, ***35***
Malayan flying ***150*** (*see* colugo, Malayan)
Mayotte 27, ***34***
Milne-Edward's sportive ***35***
Mongoose *24*, *26*, 26, 27, ***34***
Northern sportive ***35***
Philippine flying ***150*** (*see* colugo, Philippine)
Philippine gliding ***150*** (*see* colugo, Philippine)
Red-bellied ***34***
Red-fronted 27, ***34***
Red-ruffed ***35***
Red-tailed sportive ***35***
Ring-tailed 25, 26, *27*, 27, *28-29*, ***34***
Ruffed 25, 26, 28, ***34***
Sanford's ***34***
Sportive 24, *25*, 25, 26, 28, ***35***
Weasel ***35***
Western gentle ***35***
White-collared ***34***
White-footed sportive ***35***
White-fronted *25*, ***34***
Woolly 32, 33, ***35***
see also lemurs
Lemur *12*, 24, 26, 28
catta ***34*** (*see* lemur, Ring-tailed)
coronatus ***34*** (*see* lemur, Crowned)
fulvus ***34***
fulvus albifrons ***34*** (*see* lemur, White-fronted)
fulvus albocollaris ***34***
fulvus collaris ***34*** (*see* lemur, Collared)
fulvus fulvus ***34*** (*see* lemur, Brown)
fulvus mayottensis ***34*** (*see* lemur, Mayotte)
fulvus rufus ***34*** (*see* lemur, Red-fronted)
fulvus sanfordi ***34***
macaco ***34*** (*see* lemur, Black; *see also Lemur fulvus)*
mongoz ***34*** (*see* lemur, Mongoose)
rubriventer ***34***
lemurids (Lemuridae) 22, **24-28**
Lemurinae (typical lemurs) 24-25, 27-28
lemurs *10*, (26)
dwarf 22, **30-31**
gentle 24, 25, 26, 28
leaping 22, **31-33**
mouse 22, **30-31**
true 24, 26, 28
typical 24-25, 27-28
see also lemur
Leontopithecus
rosalia ***51*** (*see* tamarin, Lion)
rosalia chrysomelas (Golden-headed tamarin) 49, ***51***, 55
rosalia chrysopygus (Golden-rumped tamarin) ***51***, 55
rosalia rosalia (Golden lion tamarin) ***51***, 54-55
Lepilemur
mustelinus ***35*** (*see* lemur, Sportive)
mustelinus dorsalis ***35***
mustelinus edwardsi ***35***
mustelinus leucopus ***35***
mustelinus mustelinus ***35***
mustelinus ruficaudatus ***35***
mustelinus septentrionalis ***35***
Lepilemurinae 24
locomotion
in apes 116
in aye-ayes (31)
in bush babies 37, *38-39*
in cebid monkeys 57, *58*
in flying lemurs 150, 151
in gibbons 118
in gorillas 136
in indri 32
in Lar gibbons *118-119*
in leaping lemurs 31
in Lemuridae 24-25
in orang-utans 132, *133*, *135*
in pottos and lorises 37, *39*
in primates *10*, 15
in sub-fossil lemurs (26)
in tarsiers *43*
in typical monkeys 75-76
loris 23, 37, 38, 41
Slender *37*, *39*, ***41***
Slow *39*, ***41***
Loris
tardigradus ***41*** (*see* loris, Slender)
lorisids (Lorisidae) 23, **36-41**
Lorisinae 37
Lyonogale
dorsalis ***148***
tana ***148*** (*see* tree shrew, Terrestrial)

M

Macaca 78
arctoides ***88*** (*see* macaque, Stump-tailed)
assamensis ***88***
cyclopis ***88***
fascicularis ***88*** (*see* macaque, Crab-eating)
fuscata ***89*** (*see* macaque, Japanese)
maura ***89*** (*see* macaque, Moor)
mulatta ***89*** (*see* macaque, Rhesus)
nemestrina ***89*** (*see* macaque, Pig-tailed)
nigra ***89***
radiata ***89*** (*see* macaque, Bonnet)
silenus ***89*** (*see* macaque, Lion-tailed)
sinica ***89*** (*see* macaque, Toque)
sylvanus ***89*** (*see* macaque, Barbary)
thibetana ***89*** (*see* macaque, Père David's)
tonkeana ***89***
macaque *13*, 45, 75, 76-77, 78, **78-79**, 84
Assamese ***88***
Barbary *78*, 78, **89**, (92-93), *93*
Bear ***88*** (*see* macaque, Stump-tailed)
Bonnet *78*, 79, *82-83*, ***89***
Celebes ***89***
Crab-eating *79*, 79, *80-81*, ***88***
Formosan rock ***88***
Japanese 78, 80, ***89***, *90*
Lion-tailed 79, ***89***
Long-tailed ***88*** (*see* macaque, Crab-eating)
Moor *79*, ***89***
Père David's 78, ***89***
Pig-tailed *79*, 79, ***89***
Rhesus 74, 78, *79*, 82, 83, ***89***, (90-91), *90-91*
Stump-tailed 78, *79*, 79, ***88***
Taiwan (Formosan rock macaque) ***88***
Tibetan stump-tailed ***89*** (*see* macaque, Père David's)
Tonkean ***89***
Toque 20, 79, ***89***
mandrill *14*, *44*, 45, *76*, *77*, 79, ***89***
Mandrillus—*see Papio*
manes
Lion tamarin 54-55
mangabey 45, 75, 76-77, **80**, 81, 83, 84, (94-95)
Agile 80, ***86***
Black ***86***
Collared ***86*** (*see* mangabey, White)
Gray-cheeked 80, *84*, ***86***, (94-95), *94-95*
Red-capped ***86*** (*see* mangabey, White)
Sooty *85*, ***86*** (*see also* mangabey, White)
Tana River ***86*** (*see* mangabey, Agile)
White *74*, *85*, ***86***
marmoset 44-45, **46-49**, (48)

Bare-ear 48, **50** (*see also* marmoset, Black-tailed; marmoset, Silvery)
Black tufted-ear *47*, **50**
Black-eared **50** (*see* marmoset, Black tufted-ear)
Black-tailed *47*, **50** (*see also* marmoset, Bare-ear)
Buffy tufted-ear *47*, **50**
Buffy-headed *48-49*, **50**
Common **50**
Geoffroy's tufted-ear **50**
Pygmy *11*, 46, *47*, 48, 49, **50**
Silvery *47*, **50** (*see also* marmoset, Bare-ear)
Tassel-ear *47*, 48, **50**
mating
in chimpanzees 128-129
in tree shrews *146-147*
matrilines
baboon *20*, 20
Barbary macaque 92
macaque *20*, 20
Olive baboon (96-97)
typical monkey **83-84**
meat
gorilla 143
monkey 82
Megaladapinae 24
Megaladapis (26), *26*
Mesopithecus 102
Microcebus
coquereli **35** (*see* lemur, Coquerel's mouse)
murinus **35** (*see* lemur Gray lesser mouse)
rufus **35** (*see* lemur, Brown lesser mouse)
migration
in chimpanzees 129-130
Miopithecus
talapoin **88** (*see* monkey, Talapoin)
monkey
Allen's swamp 80, *84*, **88**
Barbe's leaf 103, ***111***
Biet's snub-nosed 103, (109), ***110***
Black snub-nosed (109), ***110*** (*see* monkey, Biet's snub-nosed)
Black spider **65**
Black-handed spider *57*, *58*, **65**
Blue 83, 84, ***88***, (100-101), *101*, 115
Bonneted leaf ***111*** (*see* monkey, Capped leaf)
Brelich's snub-nosed 103, (109), ***110***
Brown-headed spider **65**
Campbell's **88**
Capped leaf 107, ***111***
Cochin China ***110*** (*see* douc, Red-shanked)
Common woolly **65** (*see* monkey Humboldt's woolly)
Coppertail **86** (*see* monkey, Redtail)
De Brazza's 45, 80, 82, **88**
Delacour's black leaf (White-rumped black leaf monkey) ***111***
Dent's (Wolf's monkey) **88**
Diana *10*, **88**
Dollman's snub-nosed ***110*** (*see* monkey, Tonkin snub-nosed)
Dryas **88**
Dusky leaf 103, 107, *108*, ***111***
Ebony leaf ***111***
Francois's black leaf (White-sideburned black leaf monkey) ***111***
Goeldi's 46-48, *47*, 49, **51**
Golden 106, 108, (109), ***110*** (*see also* monkey, Blue)
Golden leaf 103, *104*, 107, ***111***
Golden snub-nosed ***110*** (*see* monkey Golden)
Gray-headed black leaf ***111*** (*see* monkey, Hooded black leaf)
Greater white-nosed (Spot-nosed monkey) **88**
Green **86** (*see* vervet)
Hamlyn's (Owl-faced monkey) **87**
Hendee's woolly **65**
Hooded black leaf 103, 107, 108, ***111***
Humboldt's woolly 58, *62-63*, **65**
Hussar **88** (*see* monkey, Patas)
Leonine black leaf ***111*** (*see* monkey, Hooded black leaf)
Lesser spot-nosed **87**
Lesser white-nosed (Lesser spot-nosed monkey) **87**
L'Hoest's 80, **87**
Long-haired spider **65**
Military **88** (*see* monkey, Patas)
Mona **88**
Moor leaf (Ebony leaf monkey) ***111***
Moustached *84*, **86**
Negro leaf (Ebony leaf monkey) ***111***
Night 45, 56, 58, *59*, 59, 60, 61, ***62***, (68-69), *68-69*
Orange snub-nosed ***110*** (*see* monkey, Golden)
Owl **62** (*see* monkey, Night)
Owl-faced **87**
Patas 75, 76, 77, 80, *81*, 81, 82, 83, 84, *85*, 85, ***88***
Pied black leaf (White-rumped black leaf monkey) ***111***
Pig-tailed snub-nosed 107, 108, ***110***
Proboscis *103*, 104, 106, 107, ***110***
Purple-faced leaf 107, ***111***, 114
Red colobus *14*
Red-bellied **87**
Red-eared **87**
Redtail 84, **86**, (100-101), *100-101*, 114-115
Rhesus **89** (*see* macaque, Rhesus)
Roxellane's snub-nosed ***110*** (*see* monkey, Golden)
Samango **88** (*see* monkey, Blue)
Savanna **86** (*see* vervet)
Silver **88** (*see* monkey, Blue)
Silvered leaf 107, 108, ***111***, 114
Smoky woolly **65** (*see* monkey, Humboldt's woolly)
Spectacled leaf ***111*** (*see* monkey, Dusky leaf)
Spot-nosed **88**
Squirrel 56, 57, 58, *58*, 59, 60, 61, ***64***, (66), *66-67*
Sykes' **88** (*see* monkey, Blue)
Talapoin 80, 81, 82, 83, 84, *85*, 85, ***88***
Tonkin snub-nosed 103, ***110***
Vervet **86** (*see* vervet)
White-headed black leaf ***111***
White-rumped black leaf ***111***
White-shoulder-haired snub-nosed (109), ***110*** (*see* monkey, Brelich's snub-nosed)
White-sideburned black leaf ***111***
Widow (Yellow-handed titi) **62**
Wolf's **88**
Woolly spider **65** (*see* muriqui)
Yellow-tailed woolly (Hendee's woolly monkey) **65**
see also monkeys
monkeys **44-109**
cebid 45, **56-61** (*see* capuchin)
colobine 45, 72, **102-109** (*see* colobus)
douc—*see* douc
howler—*see* howler
leaf 45, 72, **102-109**
New World 44-45
Old World 44, 45
spider 45, 59, 60, 61
squirrel 45, 57
titi—*see* titi
typical 45, **74-85**
woolly 45, 59, 61
see also monkey
monogamy
in cebid monkeys 58
in colobine monkeys 107
in Kloss gibbons 125
in primates 19, 20
in titi monkeys (70-71)
mother–infant bond
in primates 20
in typical monkeys 83
see also matrilines
muriqui 59, *61*, **65**

N

Nasalis 102
concolor ***110*** (*see* monkey, Pig-tailed snub-nosed)
larvatus ***110*** (*see* monkey, Proboscis)
nests
gorilla 138
mouse lemur 30
Ngok 41
Nogkoue 41
Nomascus 121
nostrils
monkey 44, 45
Nycticebus
coucang **41** (*see* loris, Slow)

O

orang-utan *13*, *116*, 116, *117*, 117, ***132***, **132-135**, *132-135*, (134)
Otolemur
crassicaudatus 40

P

Palaeopropithecus (26)
Pan
paniscus **126** (*see* chimpanzee, Pygmy)
troglodytes **126** (*see* chimpanzee, Common)
Papio
cynocephalus **89** (*see* baboon, Savanna)
hamadryas **89** (*see* baboon, Hamadryas)
leucophaeus **89** (*see* drill)
papio **89** (*see* baboon, Guinea)
sphinx **89** (*see* mandrill)
parking
in Lesser bush babies *41*
in pottos and lorises 38
in Ruffed lemurs 28
in tarsiers 42
parks and reserves
and Golden monkeys (109)
and gorillas 143
and Lion tamarins 55
and orang-utans 135
patrilines
ub Hamadryas baboon 99
Perodicticus
potto **41** (*see* potto)
pets
Lion tamarin 54
lorisid 41
orang-utan 134-135
tarsier 43
Phaner
furcifer **35** (*see lemur, Fork-crowned dwarf)*
pheremones
in marmosets and tamarins 48
Pithecia
albicans **64**
hirsuta **64**
monachus **64**
pithecia **64** (*see* saki, Gulanan)
Plagiomenidae 150
plantigrade gait 75
platagium
in flying lemus 150, *151*, 151, *152*
platyrrhines 44, 45
play
in cebid monkeys 61
in gorillas *141*
in Squirrel monkeys (66), 66
polygynous mating—*see* harems
Pongidae 116
Pongo
pygmaeus **132** (*see* orang-utan)
posture
in chimpanzees *10*
in Diana monkeys *10*
in gibbons *10*, *119*
in gorillas *10*
in lemurs *10*
in primates *10*, 14
in tarsiers 10
in typical monkeys 75-76
potto *14*, 37-38, 38, *39*, **41**
Golden **41** (*see* angwantibo)
Presbytis 12, 102
comata ***110*** (*see* sureli, Grizzled)
femoralis ***110*** (*see* sureli, Banded)
frontata ***110***
melalophos ***110*** (*see* sureli, Mitered)
potenziani ***110*** (*see* sureli, Mentawi Islands)
rubicunda ***111*** (*see* sureli, Maroon)
siamensis ***110*** (*see* sureli, Pale-thighed)
primates
higher 10, **44-151**
lower—*see* prosimians
Procolobus
badius ***112*** (*see* colobus, Guinea forest red)
pennantii ***112*** (*see* colobus, Pennant's red)
preussi ***112***
rufomitratus ***112***
verus ***112*** (*see* colobus, Olive)
Propithecus
diadema **35** (*see* sifaka, Diadem)
verreauxi **35** (*see* sifaka, Verroux's)
prosimians (Prosimii) 10, 13, **22-43**
protection
of Golden monkeys (109)
of indri and sifakas 33
pseudo-estrous behavior
in Hanuman langurs 115
Ptilocercinae 148
Ptilocercus
lowii ***148*** (*see* tree shrew, Pen-tailed)
Pygathrix 102
avunculus ***110*** (*see* monkey, Tonkin snub-nosed)
bieti ***110*** (*see* monkey, Biet's snub-nosed)
brelichi ***110*** (*see* monkey, Brelich's snub-nosed)
nemaeus ***110*** (*see* douc, Red-shanked)
nigripes ***110*** (*see* douc, Black-shanked)
roxellana ***110*** (*see* monkey, Golden)

R

reintroduction
of chimpanzees *129*
of Golden lion tamarins 55
of orang-utans 135
reproductive strategy
in orang-utans 133
reproductive success
in gorillas 140
in primates 20
in Redtail–Blue monkey hybrids 101
research, medical
and typical monkeys 74, 85
reserves—*see* parks and reserves

S

saggital crest
in gorillas *116*, 137
in orang-utans *117*
Saguinus
bicolor **50**
fuscicollis **50** (*see* tamarin, Saddle-back)
geoffroyi **51** (*see* tamarin, Geoffroy's)
imperator **50** (*see* tamarin, Emperor)
inustus **50** (*see* tamarin, Mottle-faced)
labiatus **50** (*see* tamarin, Red-chested moustached)
leucopus **51** (*see* tamarin, Silvery-brown bare-face)
midas **50** (*see* marmoset, Red-handed)
midas niger 48
mystax **50** (*see* marmoset, Black-chested moustached)
nigricollis **50**
oedipus **51** (*see* marmoset, Cotton-top)
Saimiri
oerstedii—*see S. sciureus*
sciureus **64** (*see* monkey, Squirrel)
saki 45,58
Bearded **64**
Black (Bearded saki) **64**
Black-bearded **64**
Buffy **64**
Guianan **64** (*see* saki, White-faced)
Monk (Red-bearded saki) **64**
Red-bearded **64**
White-faced *45*, 58, *60*, **64**
White-nosed **64**
see also sakis
sakis
bearded 57, 59
see also saki
samango **88** (*see* monkey, Blue)
Scandentia 144
scent marking
in bush babies 41
in Geoffroy's tamarins *47*
in indri and sifakas 33
in lemurs 25
in Lion tamarins 49
in marmosets 49
in primates 21
in Red-chested moustached tamarins *47*
in Saddle-back tamarins *47*
in tree shrews 149
Semnopithecus 104, 106
auratus ***111***
barbei ***111*** (*see* monkey, Barbe's leaf)
cristatus ***111*** (*see* monkey, Silvered leaf)
delacouri ***111***
entellus ***111*** (*see* langur, Hanuman)
francoisi ***111***

geei ***111*** (*see* monkey, Golden leaf)
hypoleucus ***111***
johnii ***111*** (*see* monkey, Hooded black leaf)
leucocephalus ***111***
obscurus ***111*** (*see* monkey, Dusky leaf)
pileatus ***111*** (*see* monkey, Capped leaf)
vetulus ***111*** (*see* monkey, Purple-faced leaf)
sexing
guenon 78
lorisid 37
sexual coloring
in typical monkeys 78
in vervets *81*
see also swellings, sexual
siamang *13*, 116, *117*, 117, ***120***, *120*, 121, *122-123*, 122, 123
Dwarf ***120*** (*see* gibbon, Kloss)
sifaka 31
Diadem 33, ***35***
Verroux's 32, *33*, 33, ***35***
silverback gorillas *140-141*, 140-141
simakobu ***110*** (*see* monkey, Pig-tailed snub-nosed)
simpai ***110*** (*see* sureli, Mitered)
singing
in gibbons 122-123
in indri 33
in Kloss gibbons (124-125), *124-125*
in titis 71
see also vocalizations
Sivapithecus 132
size
and higher primates *11*, *116*
and home ranges 16
and polygynous breeding 20, 142
skins—*see* furs
social behavior
in apes 116-117
in Brown capuchins (60)
in bush babies 38-41
in cebid monkeys 58-61
in chimpanzees 129-130, (130-131), *130-131*
in colobine monkeys 106-108
in Gelada baboons *18*
in gorillas *140-141*, 140-142
in Gray-cheeked mangabeys 94, 95
in Hamadryas baboons *18*, (98-99), *98-99*
in Himalayan Rhesus macaques 90-91
in Lemuridae 27-28
in marmosets and tamarins 49
in mouse and dwarf lemurs 31
in Olive baboons (96-97), *96-97*
in primates *18-22*, 18-22
in titis (70-71), *70*
spacing behavior
in Gray-cheeked mangabeys 95
in Mantled howlers 73
in orang-utans (134)
sublingua
in lorisids 36-37
sureli 103, 104, 107, 108
Banded 108, ***110***
Black-crested ***110*** (*see* sureli, Mitered)
Gray ***110*** (*see* sureli, Grizzled)
Grizzled 103, 107, ***110***
Maroon 103, ***111***
Mentawai Islands 103, 107, ***110***
Mitered 103, ***110***
Pale-thighed 104, 107, ***110***
Red-bellied ***110*** (*see* sureli, Mentawai Islands)
Sunda Island ***110*** (*see* sureli, Grizzled)
White-fronted ***110***
swellings, sexual
in baboons 77
in chimpanzees 126
in guenons 77
in mangabeys 77
in monkeys and apes 20
in Patas monkeys 78
in Rhesus monkeys 78

T

tails
prehensile 44, 57, *57*, 57, *58*
typical monkey 76
talapoin ***88*** (*see* monkey, Talapoin)
tamarin *13*, 44-45, **46-49**
Black-and-red (Black mantle tamarin) ***50***
Black-chested moustached *46*, ***50***
Black-mantle ***50***
Cotton-top *46*, 48, ***51***, *51*
Emperor *46*, 49, ***50***, (52-53), *52*
Geoffroy's *47*, 48, ***50***
Golden lion ***50***, (54-55), *54-55* (*see also* tamarin, Lion)
Golden-headed ***50***, 55 (*see also* tamarin, Lion)
Golden-rumped *47*, ***51***, 55 (*see also* tamarin, Lion)
Lion *47*, 48, 49, ***51***, (54-55), *54-55*
Midas (Red-handed tamarin) 48, ***50***
Mottle-faced *47*, ***50***
Moustached ***50*** (*see* tamarin, Black-chested moustached)
Pied bare-face ***50-51***
Red-chested moustached *47*, ***50***
Red-handed (Midas tamarin) 48, ***50***
Saddle-back *47*, *49*, 49, ***50***, (52-53), *53*
Silvery-brown bare-face (White-footed tamarin) 45, 48, ***51***
White-footed (Silvery-brown bare-face tamarin) 45, 48, ***51***
White-lipped ***50*** (*see* tamarin, Red-chested moustached)
tapetum lucidum
in angwantibos *40*
in lower primates 26, 30
tarsier *10*, *22*, 23, **42-43**
Philippine *23*, ***42***, 42, *43*
Spectral ***42***, 42, *43*
Sulawesi ***42*** (*see* tarsier, Spectral)
Western ***42***, 42, *43*
Tarsius
bancanus ***42*** (*see* tarsier, Western)
pumilus—*see T. spectrum*
spectrum ***42*** (*see* tarsier, Spectral)
syrichta ***42*** (*see* tarsier, Philippine)
territorial behavior
in apes 117
in bush babies 38-40
in colobine monkeys 107
in gibbons 122, 123
in indri 33
in Kloss gibbons (124-125)
in orang-utans 134
in Saddle-back and Emperor tamarins (52-53)
in titis 71
Theropithecus
gelada ***89*** (*see* gelada)
threat behavior
in Black-tailed marmosets *47*
in cebid monkeys 61
in chimpanzees *21*
in Goeldi's monkeys *47*
in gorillas 141
in Hamadryas baboons 99
in macaques *78-79*
in marmosets 49
in Moustached monkeys *84*
thumb
colobine monkey *102*, 102
titi 45, 57, 59, 61, (70-71), *70*
Collared ***62*** (*see* titi, Yellow-handed)
Dusky 58, ***62***, *70-71*, 70, 71
Masked ***64***, 70
Red ***62*** (*see* titi, Dusky)
White-handed ***62*** (*see* titi, Yellow-handed)
Yellow-handed ***62***, 70, 71
tool use
in chimpanzees (128), *128*
tooth-combs
dwarf lemur 31
lorisid 36
prosimian *12*, *22-23*
toxic plants
and colobine monkeys 104
tree shrew **144-149**
Belanger's 146, ***148***, 149
Bornean smooth-tailed (Southern smooth-tailed tree shrew) ***148***
Common *144*, *145*, ***148***, *149*, 149
Indian *146-147*, ***148***
Indonesian ***148***
Javan (Indonesian tree shrew) ***148***
Long-footed ***148***
Montane ***148***
Mountain (Montain tree shrew) ***148***
Nicobar ***148***
Northern smooth-tailed *145*, ***148***
Painted ***148***, 148
Palawan ***148***
Pen-tailed 144-146, *145*, 146, ***148***, 149
Philippine *144*, *145*, ***148***
Pygmy *144*, *145*, 146, ***148***
Rufous-tailed ***148***
Slender ***148***
Southern smooth-tailed ***148***
Striped ***148***
Terrestrial *144*, *145*, 146, ***148***, 149
troops
all-male 107, 114
cebid monkey 59
colobine monkey 106-107, 108
Golden monkey (109)
guereza *106-107*
Hamadryas baboon *98*, 98-99
Hanuman langur *114*, (114-115)
multi-male 19, 84-85
Olive baboon (96-97), *97*
Redtail monkey *100-101*
typical monkey **83-85**
vervet *86-87*
Tupaia 144
belangeri ***148*** (*see* tree shrew, Belanger's)
glis ***148*** (*see* tree shrew, Common)
gracilis ***148***
javanica ***148***
longipes ***148***
minor ***148*** (*see* tree shrew, Pygmy)
montana ***148***
nicobarica ***148***
palawanensis ***148***
picta ***148***
splendidula ***148***
Tupaiinae 148

U

uakari 45, 57-58, 59, 60
Black ***65***
Black-headed (Black uakari) ***65***
Red *56*, *58*, ***65***
White ***65*** (*see* uakari, Red)
underjet
in colobine monkeys 102
Urogale
everetti ***148*** (*see* tree shrew, Philippine)

V

Varecia
variegata ***34*** (*see* lemur, Ruffed)
variegata rubra ***35***
variegata variegata ***34***
vervet *74*, 78, 80, *81*, *82*, 82, 83, 84, ***86***, *86-87*
vocalizations
in cebid monkeys 61
in colobine monkeys 107-108
in gibbons 118, 122-123
in indri and sifakas 33
in Kloss gibbons (124-125), *124-125*
in mangabeys (94-95), *94-95*
in Mantled howlers *72*, 72, 73
in Night monkeys 68
in orang-utans (134)
in siamangs *122-123*
in tarsiers 42, 43
in titis 71
in tree shrews 149
in typical monkeys 84
vulnerability
of cebid monkeys 61
of chimpanzees 130
of colobine monkeys 108
of orang-utans 135

W

Wallace's line
and tarsiers 42
wanderoo ***138*** (*see* monkey, Purple-faced leaf)
whoop-gobbling
in mangabeys (94-95), *94-95*

Picture Acknowledgments

Key *t* top. *b* bottom. *c* center. *l* left. *r* right.

Abbreviations A Ardea. AN Agence Nature. BC Bruce Coleman Ltd. GF George Frame. J Jacana.

Cover BC, Norman Myers. 1 BC, Gunter Ziesler. 2–3 NHPA, Ivan Polunin. 4–5 BC, Gunter Ziesler. 6–7 Zefa, Hanumantha. 8–9 Zefa, W. H. Müller. 10*t* BC. 10*b* P. Veit. 15 A. 16–17 P. Veit. 18 R. Dunbar. 21 Frithfoto. 22 S. K. Bearder. 23 A. 26 J. Visser. 27 Natural Science Photos. 28–29 BC. 30 J. Visser. 31*t* Bob Martin. 31*b*, 32 A. 32–33, 33 J. Visser. 36 A. 37*t* BC. 37*b* S. K. Bearder. 40 P. Morris. 41 Bob Martin. 44 A. Henley. 48–49, 49, 51 BC. 52*l*, 52–53 C. Janson. 52*r* Rod Williams. 54–55 BC. 55 A. F. Coimbra-Filho. 56 BC. 57*t* Leonard Lee Rue III. 57*b* AN. 60*t* Rod Williams. 60*b* C. Janson. 61 Andy Young. 62–63 A. 67 BC. 68 C. Janson. 69 J. 70–71 BC. 71 C. Janson. 72–73 J. 74*t* A. Henley. 74*b* AN. 75 BC. 80*t* A. 80*b* Anthro-Photo. 81*t* Dawn Starin. 81*b* BC. 82*t* Eric and David Hosking. 82–83 A. 84 J. 86–87 AN. 90 A. 90–91 BC. 92 Robert Ho. 93 J. E. Fa. 94–95 Lysa Leland. 96 William Ervin, Natural Imagery. 96–97 GF. 97 Anthro-Photo, B. Smuts. 98, 98–99 H. Kummer. 102 Dawn Starin. 103 Rod Williams. 104 A. 105 J. MacKinnon. 106 J. M. Bishop. 106–107 BC. 108 A. 109 J. 113 J. M. Bishop. 114*t* A. 114*b*, 115 Anthro-Photo. 117 P. Morris. 118–119 J. MacKinnon. 119 A. Henley. 122*t* J. MacKinnon. 122*b* J. 123 BC. 126 J. 127*t* BC. 127*b* A. 128 BC. 129*t* A. Henley. 129*b* BC. 132 J. 133 J. MacKinnon. 134 Survival Anglia. 135 J. MacKinnon. 136*t* A. 136*b* BC. 137 P. Veit. 138–139 BC. 140, 141 P. Veit. 142*t* BC. 142*cl*, 142*bl* K. J. Stewart. 142*br* Syndication International. 143 P. Veit. 146 A. 146–147 BC. 147 Bob Martin. 149 BC. 150 Natural History Photographic Agency. 150–151 BC. 151 J. MacKinnon.

Artwork

All artwork © Priscilla Barrett unless stated otherwise below.
Abbreviations JF John Fuller. SD Simon Driver. AEM Anne-Elise Martin.

12, 13, 14, 18, 19, 20 SD. 22 AEM. 26 JF. 44 AEM. 48, 98 SD. 116 AEM. 120 SD. 144 AEM. 147 SD. Maps and scale drawings SD.